BIOTECHNOLOGY IN AGRICULTURE, INDUSTRY AND MEDICINE

SOYBEANS: A PEROXIDASE SOURCE FOR THE BIOTREATMENT OF EFFLUENTS

BIOTECHNOLOGY IN AGRICULTURE, INDUSTRY AND MEDICINE

Additional books in this series can be found on Nova's website under the Series tab.

Additional E-books in this series can be found on Nova's website under the E-book tab.

BIOTECHNOLOGY IN AGRICULTURE, INDUSTRY AND MEDICINE

SOYBEANS: A PEROXIDASE SOURCE FOR THE BIOTREATMENT OF EFFLUENTS

J.L. GOMEZ
M. GOMEZ
AND
M.D. MURCIA

Nova Science Publishers, Inc.
New York

Copyright © 2011 by Nova Science Publishers, Inc.

All rights reserved. No part of this book may be reproduced, stored in a retrieval system or transmitted in any form or by any means: electronic, electrostatic, magnetic, tape, mechanical photocopying, recording or otherwise without the written permission of the Publisher.

For permission to use material from this book please contact us:
Telephone 631-231-7269; Fax 631-231-8175
Web Site: http://www.novapublishers.com

NOTICE TO THE READER

The Publisher has taken reasonable care in the preparation of this book, but makes no expressed or implied warranty of any kind and assumes no responsibility for any errors or omissions. No liability is assumed for incidental or consequential damages in connection with or arising out of information contained in this book. The Publisher shall not be liable for any special, consequential, or exemplary damages resulting, in whole or in part, from the readers' use of, or reliance upon, this material. Any parts of this book based on government reports are so indicated and copyright is claimed for those parts to the extent applicable to compilations of such works.

Independent verification should be sought for any data, advice or recommendations contained in this book. In addition, no responsibility is assumed by the publisher for any injury and/or damage to persons or property arising from any methods, products, instructions, ideas or otherwise contained in this publication.

This publication is designed to provide accurate and authoritative information with regard to the subject matter covered herein. It is sold with the clear understanding that the Publisher is not engaged in rendering legal or any other professional services. If legal or any other expert assistance is required, the services of a competent person should be sought. FROM A DECLARATION OF PARTICIPANTS JOINTLY ADOPTED BY A COMMITTEE OF THE AMERICAN BAR ASSOCIATION AND A COMMITTEE OF PUBLISHERS.

Additional color graphics may be available in the e-book version of this book.

LIBRARY OF CONGRESS CATALOGING-IN-PUBLICATION DATA

Gsmez, J. L., 1956-
 Soybeans, a peroxidase source for the biotreatment of effluents / J.L. Gsmez, M. Gsmez, M.D. Murcia.
   p. cm.
 Includes index.
 ISBN 978-1-61122-088-9 (softcover)
 1. Sewage--Purification--Biological treatment . 2. Peroxidase--Industrial applications. I. Gsmez, M., 1960- II. Murcia, M. D., 1955- III. Title.
 TD755.G66 2010
 628.3'5--dc22
                                                                              2010036160

Published by Nova Science Publishers, Inc. † New York

CONTENTS

Preface		vii
Introduction		1
Chapter 1	Pollutants from Industrial Effluents that Have Been Successfully Degraded by SBP	3
Chapter 2	Treatment of Effluent Using Enzymes	7
Chapter 3	Peroxidases: Properties and Main Sources	11
Chapter 4	Soybean Peroxidase	15
Chapter 5	Removal of Phenolic Compounds with SBP	19
Chapter 6	Removal of Non-Phenolic Compounds with SBP	37
Chapter 7	Conclusion	41
References		43
Index		57

PREFACE

Nowadays the massive industrialization of society has to deal with the efficient utilization and recycling of natural resources to achieve modern industrial manufacturing processes. Many industrial wastewater streams, mainly from producing polymers, resins coatings, paints, dyes, agrochemicals, pharmaceuticals and oil refining are being discharged to the environment. Because of their toxicity level, higher than permitted, extensive research in economic and efficient technological treatments has been done recently without reaching a definitive solution.

Among the numerous applications of soybean, its use as a source of soybean peroxidase is of particular interest in the area of bioremediation. Peroxidases are oxidoreductase enzymes with a wide variety of substrate specificity, that catalyze, in the presence of hydrogen peroxide, the oxidative polymerization of different pollutants widely found in industrial effluents, such as phenolic compounds, polycyclic aromatic hydrocarbons (PAHs), aromatic amines, non-phenolic aromatics and dyes. In general, these pollutants are highly toxic for both human beings and the environment and they must be removed from wastewater before they are discharged into the environment. The use of enzymes is presented as an alternative method for the industrial wastewater treatment when conventional physical, chemical and biological methods such as adsorption in carbon active, chemical oxidation and microorganism treatment may be ineffective.

Enzymes present several advantages, mainly they are easy to handle and store, operate within a wide range of conditions, present high specificity towards the targeted compounds and minimum environmental impact. In particular, in the presence of peroxidases, the previously mentioned pollutants are oxidised forming free radicals that subsequently polymerize until the formation of high molecular weight products, being the final products insoluble polymers that can be easily precipitated from the wastewater without causing environmental problems. Soybean has the additional advantage of being a cheap and abundant source of peroxidase, easily extracted from soybean seed hulls that are a by-product in the food industry, and capable of maintaining its enzymatic activity over a wide range of temperature and pH, which explains its increasing use over the last years in the biotreatment of effluents. In order to decrease the systems costs one important area for future research will be the study of the direct use of soybean seed hulls without any extraction of the enzyme.

INTRODUCTION

Massive discharges of industrial wastewater without appropriate treatment commonly occur in many developing countries, causing severe damage to both human beings and the environment. As a result, over the last decades, environmental and health concerns have arisen and bioremediation has become an area of great interest for all modern industrial processes generating contaminated effluents. In this sense, numerous efforts have been focused on the search of convenient and cost effective wastewater treatments.

Among the different wastewater treatments, enzymatic treatments have proven to be good alternatives to traditional biological treatments, showing some important advantages like their applicability to a wide range of compounds, catalytic ability over wide ranges of pH, temperature and substrate concentration, minimal environmental impact, etc.

In particular, peroxidases are a type of oxidoreductases enzymes that catalyze bisubstrate reactions based on the reduction of peroxides and the oxidation of a wide variety of organic and inorganic compounds. Among these compounds, peroxidases show affinity towards different pollutants commonly found in industrial wastewater, such as phenolic compounds, polycyclic aromatic hydrocarbons, aromatic amines, non-phenolic aromatics and dyes.

Soybean peroxidase (SBP) belongs to Class III plant peroxidases family. It can present high activity and concentration in soybean seed hulls, where it is usually extracted from. And since the seed coat is a by-product of the soybean food industry, SBP has high potential of being a cost effective alternative in wastewater treatment, being its high thermostability another significant advantage.

The present work shows a review of the current state of use of soybean peroxidase in bioremediation of wastewater, and, particularly, in the removal of the following groups of pollutants: phenolic compounds, polycyclic aromatic hydrocarbons, aromatic amines, nitroaromatics, benzothiazoles and paper dyes. The chapter is divided in different sections for a detailed explanation of the most relevant aspects related to this research area, such as the following ones:

- The main characteristics and toxic effects of the different pollutants that have been successfully degraded by soybean peroxidase.
- The advantages of wastewater treatment using enzymes and, particularly, peroxidases.
- Catalytic properties and main sources of peroxidases.
- Main properties and characterization of soybean peroxidase for wastewater treatment.
- Review of the current state of research in bioremediation using SBP as catalyst for the removal of different pollutants, such as phenolics, PAHs, aromatic amines, etc. Future trends.

Chapter 1

POLLUTANTS FROM INDUSTRIAL EFFLUENTS THAT HAVE BEEN SUCCESSFULLY DEGRADED BY SBP

PHENOLIC COMPOUNDS

Phenolic compounds are a large group or aromatic compounds widely present in industrial effluents. Environmental contamination with phenolics occurs from wastewater generated by high-temperature coal conversion, petroleum refining and the manufacture of plastics, resins, textile, iron, steel and paper [1, 2].

Among the numerous groups of pollutants present in industrial effluents, phenolic compounds are widely known to be priority pollutants. They have a low biodegradability and are harmful for organisms even at ppb levels, being most of them toxic and some suspected to be carcinogens [3].

Chlorinated phenols (CPs) are particularly dangerous since an increasing number of chlorine atoms in the benzene ring raise the toxicity of CPs, their stability against decomposition, and the capability for bioaccumulation. Some of them have mutagen and carcinogenic properties. Additionally, CPs are the precursors of more dangerous ecotoxicants, dioxins [4]. They are classified as hazard chemicals [5, 6] and constitute a particular group of priority toxic pollutants listed by the US EPA in the Clean Water Act [7] and by the European Directive 2455/2001/EC [8].

POLYCYCLIC AROMATIC HYDROCARBONS

Polycyclic aromatic hydrocarbons (PAHs) are among the most common pollutants in the world. They are a group of over 100 different chemicals formed during the incomplete burning of coal, oil and gas, garbage, or other organic substances [9]. They generally exist as complex mixtures although some of them are manufactured. As pure chemicals, PAHs are colorless, white, or pale yellow-green solids. They can have a faint, pleasant odor. A few PAHs are used in medicines and to make dyes, plastics, and pesticides. Others are contained in asphalt used in road construction. They can also be found in substances such as crude oil, coal, coal tar pitch, creosote, and roofing tar. They are present in the air, water, and soil. They enter water through wastewater discharge and due to their low solubility they stick to solid particles and settle to the bottoms of lakes or rivers.

Studies in animals have shown that PAHs can cause harmful effects on skin, body fluids, and the body's system for fighting disease after both short- and long-term exposure. These effects have not been reported in people. However, The U.S. Department of Health and Human Services (DHHS) has determined that some PAHs may be carcinogens to human beings as a result of long term exposure by breathing or skin contact.

PAHs of low molecular weight are easily biodegradable [10] but those with high molecular weight (normally with four or more aromatic rings) are usually quite recalcitrant and toxic [11, 12].

AROMATIC AMINES

Aromatic amines constitute a wide group of chemicals derived from aromatic hydrocarbons, such as benzene, toluene, naphthalene or anthracene, by substituting at least one hydrogen atom for an - NH_2 amino group. They are present in the wastewater of numerous industries such as the manufacturing of dyes and other organic chemicals, the coal conversion industry, resins and plastics manufacturing or the textile industry [13].

Nearly all aromatic amines are toxic and though their effects must be considered individually most of them share some characteristics: they have been reported to be carcinogens, hemotoxicants and to cause skin damage [14, 15].

NITROAROMATICS

Nitroaromatics (NACs) are massively produced, being among the most common anthropogenic pollutants. They are used as munitions, herbicides, insecticides, pharmaceuticals, solvents and in the manufacture of dyes and plastics [16]. They can also be formed in the environment from aromatic contaminants.

Nitroaromatics are highly toxic and hazardous and some of them have been identified as human carcinogens. The list of 129 priority pollutants from the U.S. Environmental Protection Agency includes seven NACs: nitrobenzene; 2,4- and 2,6-dinitrotoluene; 2- and 4-nitrophenol; 2,4-dinitrophenol; and 4,6-dinitro-2-methylphenol [17]. Among them, dinitrotoluene is of particular interest. It is toxic by all routes (inhalation, ingestion, and dermal absorption) and it is known to be a carcinogen and a cause of anaemia, nerve and liver damage [18].

BENZOTHIAZOLES

Benzothiazoles are a wide class of compounds that result from natural processes or chemical manufacturing. They are commonly used as fungicides, herbicides, corrosion inhibitors and mainly as vulcanization accelerators in rubber production, being present in wastewaters from rubber additive manufacturing and tanneries [19].

Benzothiazoles inhibit microorganisms activity in conventional biological wastewater treatments, being most of them hardly biodegradable. They are adsorbed into cell membranes, leading to bioaccumulation [20].

Among the benzothiazoles, 2-mercaptobenzothiazole is one of the most important members, being highly recalcitrant, toxic and allergenic [21].

PAPER DYES

Dyes are coloured, ionising and aromatic organic compounds which show an affinity towards the substrate they are being applied to. These substrates are mainly textiles, leather, plastic, paper etc. As a result, dyes are commonly found in wastewaters and, once released into the receiving water bodies, can be extremely dangerous to aquatic life, as toxicants and as substances that reduce penetration of photosynthetic radiation curtailing biological activity [22].

In particular, decolorization of paper dyes is of great interest since it can allow paper recycling. Concern about recyclability of natural resources has become a main aspect of modern industrial manufacturing processes and a major difficulty in recycling paper is the problem associated with decolorizing the dyes present in the paper [23].

The most common dyes in paper industry are basic and direct dyes, which present high affinity for lignin and cellulose, respectively. Many basic dyes exhibit high aquatic toxicity. However problems are more often attributable to improper handling procedures, spill clean-up and other upsets [24]. Direct dyes are generally less toxic although they exhibit a lower fixation degree than basic dyes and as a result additives are usually required in the dye formulation, increasing the final toxicity.

Chapter 2

TREATMENT OF EFFLUENT USING ENZYMES

MAIN POLLUTANTS TREATMENTS

There is a wide variety of treatments that can be used for the removal of the previously mentioned pollutants [25]. The most relevant, classified as chemical, physical and biological treatments, are indicated in Table 1.

ADVANTAGES OF ENZYMATIC TREATMENTS

The use of enzymes in wastewater treatments has become very popular over the last decades [26]. They can act on specific contaminants and eliminate them either by precipitation or by forming new products that can be more easily removed [27]. The main advantages of enzymatic treatments [28] are listed as follows:

- Applicability to a wide range of compounds
- Low energy requirements
- High degree of substrate specificity
- High reaction rates
- Mild operation conditions
- Easy to handle processes
- Applicability to a wide range of compound concentrations

- Catalytic ability over wide ranges of pH, temperature and substrate concentration
- Appropriate to treat recalcitrant compounds
- Minimal environmental impact

Table 1. Pollutant treatments classification

CHEMICAL TREATMENTS	Non photochemical processes • Ozonation • Fenton processes (Fe^{2+}/H_2O_2) • Electrochemical oxidation • Radiolysis • Plasma processes • Ultrasounds Photochemical processes • Oxidation in sub/supercritical water • Water photolysis in UV and vacuum • UV (UV/UVV) • Ultraviolet/Hydrogen peroxide (UV/H_2O_2) • Ultraviolet/ozone (UV/O_3) • Photo-Fenton • Heterogeneous photocatalysis
PHYSICAL TREATMENTS	• Adsorption • Extraction • Membrane technology
BIOLOGICAL TREATMENTS	• Microorganisms • Enzymes • Plants (Phytoremediation)

ENZYME IMMOBILIZATION

Frequently enzymes are immobilized on different kinds of supports. Immobilized enzymes show some advantages compared with their soluble forms. They present higher operational stability and can be easily separated from the solution for reuse, which allows continuous treatment and development of cheaper processes [29]. However, immobilization of enzymes presents some risks, mainly the alteration of the enzyme conformation and the activity loss during the immobilization procedure.

There are many different methods for enzymes immobilization, such as physical adsorption, covalent bonding, crosslinking, inclusion and encapsulation. They are usually classified according to the nature of the interactions responsible of the process [30, 31] or the type of support used [32].

The use of immobilized soybean peroxidase in wastewater treatment will be in depth discussed in the following sections of this chapter.

Chapter 3

PEROXIDASES: PROPERTIES AND MAIN SOURCES

CATALYTIC CYCLE OF PEROXIDASES

Peroxidases are oxidoreductases enzymes that catalyze bisubstrate reactions based on the reduction of peroxides and the oxidation of a wide variety of organic and inorganic compounds, including phenols, aromatic amines, etc. Most peroxidases are hemoproteins and have hydrogen peroxide as common substrate [33].

The catalytic cycle of peroxidases involves distinct intermediate enzyme forms. There is a two-electron oxidation of the heme group by hydrogen peroxide that leads to the formation of an unstable intermediate known as compound I. Two successive one-electron reductions in the presence of two molecules of the electron donor substrate return the enzyme to its resting state using a second intermediate, compound II. As a result of this process, free radicals are formed that will subsequently lead to the formation of high molecular weight products that can be easily removed by precipitation due to their low solubility. This is one of the main advantages of the use of peroxidases in wastewater treatments [34]. This catalytic cycle will be described in detail in the section of kinetics of this chapter.

MAIN SOURCES OF PEROXIDASES

Peroxidase activity has been identified in plants, microorganisms, fungi and even in animals. They all belong to a family consisting of three major classes, where Class III peroxidase is the most important one.

Microbial and Fungi Peroxidases

The main microbial and fungi peroxidases used for the removal of organic pollutants and dyes are peroxidases from *Arthromyces ramosus* [35, 36], *Coprinus cinereus* [37-39], *Coprinus macrorhizus* [40, 41], chloroperoxidase from *Caldariomyces fumago* [42-45], and manganese and lignin peroxidases from the white-rot fungus *Phanerochaete chrysosporium* [46-53].

Microbial and fungi peroxidases present some important disadvantages, like their inactivation and, mainly, their low concentration in the production sources. For this reason, plant peroxidases are usually preferred.

Plant Peroxidases

Peroxidases play an important role in plants, participating in the lignification process and in the mechanism of defence in physically damaged or infected tissues. They possess a wide variety of substrate specificity and can oxidise a large number of aromatic compounds in the presence of H_2O_2 [54].

The most popular family of plant peroxidases, with more than 3000 members [55], is known as Class III plant peroxidase (POX) and consists of the secretory plant peroxidases. POXs exist as isoenzymes in individual plant species and each isoenzyme has variable amino acid sequences and expression profiles which make them appropriate for specific physiological processes [56].

Horseradish peroxidase (HRP), extracted from horseradish roots, is the most studied Class III plant peroxidase and has offered promising results regarding wastewater treatment at industrial scale. In particular, HRP has shown high affinity towards phenolic compounds and aromatic amines [57-63].

The high costs of the treatment together with enzyme inactivation are the main problems concerning the use of HRP. Trying to minimize these factors, many authors have worked to optimize reaction conditions through different methods: optimization of pH and temperature [64-66], use of additives to reduce enzyme inactivation [67-70], enzyme immobilization [71-75] or selection of the most appropriate reactor configuration [76-78].

Other plant peroxidases used in wastewater treatment, mainly in the removal of phenolic compounds, are artichoke [79], turnip [80, 81], barley [82], plant root surface peroxidases [83], etc.

Chapter 4

SOYBEAN PEROXIDASE

MAIN PROPERTIES OF SBP

Soybean peroxidase (SBP) belongs, like HRP, to Class III plant peroxidases. It can be detected in root, leaf and seed hulls of the soybean. However it is extracted from seed hulls where it presents the highest activity and can be found at high concentrations [54, 84].

SBP was first identified in 1991 as a rich source of a single peroxidase isoenzyme with a molecular weight of 37 KD [84]. Both SBP and HRP present a very similar three-dimensional structure with a sequence homology of 57%, according to the results of different spectroscopic studies [85-87]. And since the seed coat is a by-product of the soybean food industry, SBP has high potential of being a cost effective alternative to HRP in wastewater treatment, where it presents high affinity towards aromatic compounds such as PAHs or phenolics. In fact, raw soybean hulls have successfully been tested for the removal of phenolic compounds, which proves that SBP can be used without extraction or purification [88].

Together with its low cost, the main advantage of the use of SBP is its high thermostability. As a result, it has proven to be of great interest for many applications, not only as biocatalyst for the removal of pollutants from industrial effluents, but also to be used in medical diagnostic tests, replacement of formaldehyde in adhesives, abrasives and protective coatings, synthesis of phenolic resins, etc [89].

CHARACTERIZATION OF SBP FOR WASTEWATER TREATMENT

The biocatalytic properties of soybean peroxidase have been widely investigated and most authors have concluded that SBP is able to retain its catalytic properties under wide ranges of pH and at elevated temperatures [54]. In particular, thermal inactivation of SBP occurs at 90.5 °C, nine degrees above HRP and more than 25 °C above other peroxidases [89].

Wright and Nicell developed a detailed study, determining the enzymatic activity of SBP using phenol as substrate and under different conditions and periods of incubation [90]. They concluded that activity can be observed between pH 3 and 9, being pH 6.4 the optimal one. SBP incubated at 25 °C is very stable at neutral and alkaline conditions. However, rapid inactivation occurs below pH 3. Besides, acid and alkaline conditions can generate different reaction products, such as soluble oligomers of unknown nature and toxicity. On the other hand, when SBP is incubated at 80°C and pH between 5 and 10.5 the enzyme shows a thermal deactivation depending on pH. The enzyme can also be deactivated for its incubation in hydrogen peroxide. In comparison with HRP, SBP has lower catalytic activity and is slightly more sensitive to pH, but it is more stable at elevated temperatures and less susceptible to inactivation by hydrogen peroxide.

Similar studies of SBP activity and stability can be found in the literature. Geng et al. [54] report enzymatic activity between pH 2.2 and 8, with pH 6.0 as the optimal one. According to their research SBP shows three times higher activity at an elevated temperature (80 °C) at pH 6.0 when compared to the activity at room temperature. They tested different organic solvents, proving that SBP is fairly active in them, with the highest activity in the presence of 16.67 % (w/v) ethanol followed by acetone, methanol and acetonitrile. Amisha and Behere [91] have calculated a conformational stability of SBP under physiological conditions of 43.3 kJ/mol, which is much higher than the 17 kJ/mol reported for HRP. Maximum catalytic activity and conformational stability of SBP was observed at pH 5.5. They also compared the thermal stability of HRP and SBP at pH 7.0, being considerably higher for SBP (86 °C versus 74 °C).

It has been reported that after the enzymatic reaction the formed particles precipitate out of the solution entrapping the SBP and thereby reducing its activity. Many authors have demonstrated the efficiency of the use of additives to prevent the entrapment and reduce enzyme inactivation, being polyethylene glycol (PEG) the most commonly used [92, 93]. It was concluded that the effectiveness of PEG increased with molecular weight, being 35000 the most effective. PEGs with molecular weighs of 1000 and below were ineffective in protecting SBP.

Chapter 5

REMOVAL OF PHENOLIC COMPOUNDS WITH SBP

FREE SBP

Soybean peroxidase (SBP) [84] has proved to be very effective in removing phenolic compounds from wastewater [92-98]. It has been characterized and has been shown to effectively treat phenolic compounds such as, phenol, 2-chlorophenol, 3-chlorophenol, 4-chlorophenol, 2,4-dichlorophenol, and 3-methylphenol [90]. Besides, being soybean a cheap and abundant source of peroxidase, the use of SBP would help to convert a waste into a value-added product [99-100].

McEldoon et al. [97] have examined the substrate specificity of a peroxidase isolated from soybean hulls. In their study, they found that vegetable peroxidases possess wide substrate specificity than can be dramatically broadened at low pH values to include non-phenolic compounds. Taylor et al. [101] provided a limited comparison of HRP and SBP for treatment of phenols and concluded that SBP was an effective alternative. After that, Caza et al. [92] studied the removal of different phenolic compounds such as phenol, chlorinated phenols, cresols, 2,4-dichlorophenol and bisphenol A, with soybean peroxidase. The reaction parameters, optimized to achieve a removal efficiency of at least 95% were pH of 8, SBP dose without PEG of 0.20 U/mL, SBP dose with PEG of 0.15 U/mL and different [H_2O_2]/[substrate] optimum depending on the chosen substrate. The results of the study have demonstrated the applicability of using SBP for treating synthetic phenolic wastewater. The study recommended the necessity of investigating the potential use of soybean hulls instead of purchasing the enzyme from a chemical manufacturer. After the enzyme has been extracted from the hulls, they could then be used as animal feed. This would greatly reduce the amount of solid waste that is produced and would also reduce the costs. Similarly, some researchers have also suggested using crude HRP [59, 61, 102] to reduce treatment cost. In contrast to crude SBP, the amount of solid waste produced with the use of crude HRP is much greater. Another advantage of using soybeans is that they are more readily available than horseradishes.

Kenedy et al. [103] studied the removal of 2,4-dichlorophenol (2,4-DCP) with soybean peroxidase. They found that in the presence of additive PEG-3350 and PEG-8000 2,4-DCP removal efficiency of SBP increases by factor of 10 and 50 respectively compared to no PEG addition. Optimum pH for removal of 2,4-DCP without PEG was pH 8.2. The pH operating range of SBP was from 2.5 to 9.4 which is wider than reported for horseradish peroxidase. A new pH optimum of 6.2 was also found when SBP was used with PEG. By using optimum pH conditions, PEG-8000 and SBP concentration can be optimised coincidentally to minimise the amount of SBP needed while simultaneously limiting the residual PEG that might be released to the environment.

Bódalo et al. [75] have compared the efficiency of HRP and SBP in phenol removal, proving they are both quite active in a pH range of between 6.0 and 8.0, with maximal removal efficiency at neutral pH. HRP acts faster than SBP but is more susceptible to inactivation, although it is better protected by PEG.

The same authors [104] studied the removal of 4-chlorophenol using free soybean peroxidase and hydrogen peroxidase in a discontinuous tank reactor in a first step to reduce the concentration of buffer solutions used and to avoid adding polyethylene glycol. Most of the assays provided good removal percentages, which reached almost 90% for an enzyme concentration of 7.5 10^{-3} mg/cm^3 and a stoichiometric ratio of 1:1 for both substrates, when concentrations of 1.5 mmol/dm^3 or less were used.

Parra et al. [105] studied the removal of 2,4,6-trichlorophenol (TCP) with SBP. They found that the best conditions for the biodegradation in 15 min of 1mM TCP were 27 nM SBP, 1 mM hydrogen peroxide and 50 mM pH 5.0 sodium citrate buffer at 25 °C.

IMMOBILIZED SBP

As it has been previously commented the most important drawback of enzymatic methods for phenolic compounds removal is the short catalytic lifetime due to enzyme inactivation by various side reactions of the treatment process. In this regard, the use of protective additives that decrease enzyme deactivation, while being inexpensive and non-toxic, has been described [68, 106, 107].

Polyethylene glycol (PEG) was found to be one of the best additives [106, 108] and its addition has been demonstrated to be effective in diminishing HRP inactivation. However, the use of PEG in the reaction medium may have negative effects. Although it has been demonstrated that PEG is removed with the phenol polymers, Kinsley and Nicell [93] established that a great part of PEG added to the reaction medium remains in the solution after the enzymatic treatment. This impact of residual soluble PEG on downstream processes or on receiving water bodies may restrict its practical application during wastewater treatment.

Another approach which has been utilized to increase the potential use of enzymes in wastewater treatment is immobilization. For the treatment of large volumes of wastewaters, reactors containing immobilized enzymes are desirable because of the high cost of enzymes [106]. Although many supports and methods have been used for HRP immobilization very few papers have been found concerning the use of immobilized SBP for phenolic compounds removal.

Gomez et al. [109] studied the immobilization of soybean and horseradish peroxidases on glutaraldehyde-activated aminopropyl glass beads and its application to phenol removal. Glass beads were selected as immobilization matrix because of their excellent mechanical properties and because they can be modified to include a variety of functional groups. The alkylamine-support was obtained by aqueous silanization which appears to produce a monolayer of silane across the carrier surface of high hydrolytic stability, providing greater support and a sufficient degree of enzyme loading. The modified support was then activated with glutaraldehyde and covalently linked to the enzymes via available amino functions. Cross-linking was avoided by washing the support thoroughly adding the enzyme to remove all excess glutaraldehyde. They found that immobilized SBP is less susceptible to inhibition/inactivation than HRP and so provides higher conversions. The same research group [110] used the same method and support for SBP immobilization in order to study 4-chlorophenol removal. They found that SBP was less susceptible to inactivation than HRP providing higher 4-chlorophenol removal. Besides, the immobilization exerts a protective effect against the enzyme inactivation by hydrogen peroxide.

Lately, this research group has used the same immobilization procedure publishing numerous papers about phenolic compounds removal with soybean peroxidase in different types of reactors. Apart from this research group, papers using the peroxidase seed hulls entrapped within hybrid (silica sol gel/alginate) particles directly as a catalyst were found [88, 111, 112].

Gacche et al. [113] studied the removal of phenol from industrial wastewaters using immobilized soybean peroxidase on fibrous aromatic polyamide. They achieved a phenol removal percentage of 91.25%, reducing phenol concentration from 0.720 to 0.063 mg/mL after 5 hours of treatment.

Magri et al. [114] immobilized soybean peroxidase on different polyaniline polymers previously treated with glutaraldehyde. Good immobilized derivatives were obtained with 8.2 mg SBP/g support and with 82% of retained activity.

CONTINUOUS REACTORS FOR THE REMOVAL OF PHENOLIC COMPOUNDS

In most of the previous studies free enzymes were used in discontinuous tank reactors, which made them impossible to reuse and eliminate from the bulk reaction.

As mentioned above, the use of immobilized enzymes has several advantages. For instance, they can be easily separated from the reaction products and reused. They also work better in continuous processes. Continuous processes with immobilized enzymes can be carried out in different reactor configurations, although fluidized bed reactors have certain advantages over conventionally used tank and packed bed reactors. Fluidized beds are specially recommended when the substrates are viscous or contain suspended particles. The pressure drop is smaller when using fluidized bed reactors and the flow distribution through the reactor radial section is more uniform, which minimizes the formation of preferential channels. Also, since mechanical stirring is not necessary, the support particles are less damaged by abrasion. All these advantages are of special interest in the phenolic compound oxidation process using immobilized peroxidase, since most of the products formed are insoluble, while in packed bed reactors this would lead to clogging phenomena and to an increase in the pressure drop.

However, despite the many papers published on the use of fluidized bed reactors for the removal of phenolic compounds with immobilized microorganisms, very few works have used immobilized enzymes for the same purpose. Wu et al. [78] compared the behaviour of phenol removal with free peroxidase in the presence of PEG in different types of reactors: batch-mixed, continuous stirred tank reactor and plug flow. They used a previously developed model in a batch reactor to predict the process in different types of reactor system. Another work developed by Gomez et al. [115] studied the removal of 4-chlorophenol with free soybean peroxidase and hydrogen peroxide in a continuous tank reactor. They concluded that the continuous operation of a tank reactor without using buffered media provided a good degree of 4-chlorophenol elimination under all the experimental conditions assayed.

As regards published papers related to fluidized bed reactor with immobilized peroxidases, Trivedi et al. [112] tested soybean peroxidase entrapped on a hybrid silica/alginate gel, for the removal of 85% of phenol. This study concluded that the obtained particles can be used in packed bed and fluidized bed reactors. The same authors [116] used the same immobilized particles in a fluidized bed reactor obtaining phenol removal percentages of 54% with molar ratio phenol/hydrogen peroxide of 1/2.

Gomez et al. [117] used immobilized derivatives of soybean peroxidase, covalently bound to glass supports with different surface areas, in a laboratory scale fluidized bed reactor to study their viability for use in phenol removal. 80% phenol removal was achieved when derivatives immobilized on supports with the highest surface area were used.

The same authors [118] worked with the same immobilized derivative of soybean peroxidase to study phenol removal in a continuous tank reactor. Efficiency of about 80% was attained in phenol removal for the best operational conditions assayed. With the aim of scaling up the process to pilot-plant scale, the influence of the main plant operational variables, enzyme and substrate concentrations and flow rate, on the removal efficiency was studied. The continuous reactor was operated under isothermal conditions at the optimal temperature established in the literature for this system. As a first approximation to the industrial application of the process, no buffer solutions were used.

ULTRAFILTRATION CONTINUOUS TANK REACTORS

For the continuous operation of economically viable processes in enzymatic reactors, the biocatalyst must be separated from the reagents. Ultrafiltration has proved to be a very versatile separation process for biocatalysts, and the development of asymmetric membranes composed of an ultra-thin microporous layer supported by a macroporous structure allows this technique to be applied to the separation of enzymes.

The concept of ultrafiltration membrane reactors (UFMR) is based on the potential of semipermeable membranes to retain the biocatalyst but not the products formed during the reaction.

The main advantages of continuous operation with soluble enzymes are that:

- the catalyst can be homogeneously distributed, thereby avoiding transport limitations;
- almost all the native activity can be used;
- no special immobilization know-how is needed,
- while immobilization on carriers is normally expensive and time consuming.

Normally, both immobilized and soluble biocatalysts lose activity with time, and so, one additional advantage of UFMR is that fresh enzyme can be added easily to maintain productivity.

This reactor configuration has also been used for 4-chlorophenol removal. Murcia et al. [119] developed an intensive combined treatment for removing 4-chlorophenol, consisting of its oxidation with soluble soybean peroxidase and hydrogen peroxide in a continuous tank reactor, followed by purification of the effluent in an ultrafiltration membrane module in series. High 4-chlorophenol conversion percentages were reached in all cases. Although the conversion percentages in the reactor were only maintained for 40 min in some cases, they remained practically constant throughout the operational time in the permeate stream, where the membrane somehow acted as a support for successful enzyme immobilization.

DIRECT USE OF SOYBEAN SEED HULLS WITHOUT EXTRACTION

One of the advantages of soybean peroxidase versus other enzymes is the potential use of the direct soybean hulls instead of purchasing the enzyme from a chemical manufacturer. In this sense Gijzen et al. [120] observed that seed coat of soybean presented higher activity than in the soybean roots. Besides, due to the easy extraction of the enzyme from the seed coat of soybean its use looked feasible. Flock et al. [88] studied the removal of phenol and 2-chlorophenol using soybean seed hulls in the presence of hydrogen peroxide. The following conclusions were obtained from this paper:

- The direct use of soybean seed hulls reduces the biocatalyst cost, because extraction and purification steps are removed.
- The hulls act as an enzyme protector agent preventing the enzyme from adsorbing in the formed polymer.
- It has been proved that the hulls do not adsorb phenolic compounds. So their removal is only due to enzymatic reaction.
- Removal percentages of 60% were obtained in discontinuous tank reactors, whereas four sequential batch reactors achieved greater than 96% removal of phenol with a retention time of 20 min in each reactor.
- The stability of the soybean peroxidase enzyme was examined in the presence of detergents. Low concentrations of the detergents (0.1% w/v) increased the enzyme activity and higher concentrations of detergents (> 0.1% w/v and up to 20% w/v) did not activate the SBP enzyme.

Bassi et al. [111] studied the direct use of soybean seed hull extracts for the bioremediation of phenolic wastes. The crude SBP extracted from the hulls, like pure soybean peroxidase, was catalytically active in a broad range of pH and temperatures. Besides, the direct use of soybean seed hulls can reduce SBP inactivation by H_2O_2 and enhance the utilization efficiency of SBP through the slow release of the enzyme from the seed hulls. As the crude extract contains a mixture of multiple soybean proteins, soybean seed hull slurry required a higher concentration of H_2O_2 to remove the phenolic substrates than did the purified enzyme. Experiments carried out in a batch reactor removed 80% of phenol (10.6 mM), 96% of 2-chlorophenol (3.9

mM), 95% of 2,4-dichorophenol (3.1 mM) and 94 % of mixed phenol and chlorophenols. Although the results were satisfactory the authors suggest doing more studies to choose the adequate type of reactor.

Lately, the same research group [112] has continued with this research line. They studied the phenol removal using sol-gel/alginate immobilized soybean seed peroxidase in fluidized bed reactors. The main conclusions obtained from this work were the followings:

- The seed hull of soybean is a peroxidase enzyme source and can be applied for the removal of phenolic compounds.
- The alginate incorporation to the immobilized derivative is a good method to obtain particles of uniform size which are suitable for their use in fluidized bed reactors.
- The immobilized SBP can be reused for its application in discontinuous processes after a regeneration treatment.
- The immobilized derivative retains almost 90 % of its activity after 3 weeks being stored at 4 °C.

The same authors [116] studied the reaction and regeneration of immobilized SBP particles in fluidized bed reactors, proving that this reactor configuration is suitable for processes in which the biocatalyst deactivation is an important issue.

REAL WASTEWATERS

After the removal of phenolic compounds with free and immobilized soybean peroxidase in different types or reactors, discontinuous and continuous, the next step should be to try with real wastewaters instead of synthetic ones. In this sense, few real wastewater matrices have been investigated mainly due to the fluctuating pH, temperatures and pollutant concentrations, number of other compounds such as heavy metals, inorganic salts and organic compounds presented in the medium in various ratios.

Steevensz et al. [121] compared two enzymes, soybean peroxidase and laccase, to convert phenol in synthetic and refinery wastewater, with and without the influence of possible interferents. Taking into account the obtained results for soybean peroxidase it is possible to point out that the broad operating pH range would allow direct treatment of the refinery stream. Refinery samples need higher enzyme concentration (1.2 to 1.8 fold) than synthetic samples for removing 1.0 mmol/L of aqueous phenol in 3 hours. This study concluded that enzymatic treatment is suitable for phenol removal from refinery wastewater.

KINETICS OF THE ENZYMATIC REACTION

Mechanisms

Usually, the oxidation of aromatic compounds with hydrogen peroxide, catalyzed by peroxidase, has been described through the mechanism postulated by Chance-George, also known as the Dunford mechanism [122, 123], which is referenced in most of the papers found in the literature. Although the kinetic model has been validated with horseradish peroxidase, it is accepted for the rest of the peroxidases as well.

The steps of this mechanism are the following ones:

$$E + H_2O_2 \rightarrow E_1$$
$$E_1 + \Phi H_2 \rightarrow E_2 + \Phi H \cdot$$
$$E_2 + \Phi H_2 \rightarrow E + \Phi H \cdot$$
$$\Phi H \cdot + \Phi H \cdot \rightarrow H\Phi - \Phi H$$

The native enzyme, E, in the presence of hydrogen peroxide, forms a compound, E_1, called Compound I, which, in turn, accepts an aromatic compound, ΦH_2, oxidizing it and giving the free radical $\Phi H \cdot$. This radical is released to the reaction bulk and the enzyme changes to the E_2 state, Compound II, which is able to oxidize another molecule of ΦH_2, giving another free radical and returning to the native state, E, so that the cycle is closed. The overall reaction is:

$$H_2O_2 + 2\Phi H_2 \xrightarrow{E} 2\Phi H \cdot + 2H_2O$$

The above mechanism is based on the discovery of species E_1 and E_2 and the additional observation of the spontaneous transformation of E_1 in E_2, being this transformation favoured in the presence of reductive species in the reaction media.

Stoichiometric studies [122] have shown that the species E_2 only contains one of the two oxidation equivalents of the hydrogen peroxide molecule and that it is a covalent compound. In this sense, it has been proved that using CH_3OOH instead of H_2O_2 a similar reaction takes place in the first step of the above mechanism. However, the definitive structure of this species has not been well established until now [76] and is usually described by the above mentioned Chance-George mechanism. As described in the literature [122, 124, 125], in an excess of hydrogen peroxide E_2 can be oxidized to E_3, which appears as an inactive form of the enzyme:

$$E_2 + H_2O_2 \rightarrow E_3 + H_2O$$

This is not an irreversible deactivation because E_3 breaks down spontaneously to the native form [126], although with a low reaction rate:

$$E_3 \rightarrow E \cdot + O_2^-$$

Buchanan and Nicell [127] also proposed other secondary reactions, where E_3 can be reduced accepting an electron from phenol:

$$E_3 + \Phi H_2 \longrightarrow E_1 + \Phi H \cdot$$

Baynton et al. [128] observed that, for peroxide concentrations below 1 mM and in the absence of phenol, a reversible inactivation took place through formation of an intermediate compound of the enzyme, called Compound III, from which the native enzyme can be regenerated. On the other hand, in the presence of aromatic substrates, an irreversible inactivation that depends on the substrate concentration, phenol in this case, was observed, due to the reaction of the native enzyme and the intermediates Compounds I and II with phenolic radicals.

Nicell et al. [76] used horseradish peroxidase in simple and sequential continuous tank reactors to oxidize phenol and 4-chlorophenol. These authors described three enzyme inactivation pathways. Besides the formation of E_3 in excess of peroxide, they believe that a permanent inactivation, by the union of a free radical to the active centre of the enzyme, can take place, as well as an adsorption in the formed polymer. In another work [70], also with HRP and phenol, Nicell et al. established that polyethylene glycol has a protective effect on these two inactivation modes, and consequently the same authors determined the minimum amount of polyethylene glycol necessary to achieve maximum protection of the enzyme, as a function of the initial phenol concentration.

Patel et al. [129] focused their research on the oxidation of substituted phenolic compounds by the HRP Compound II. According to these authors, the HRP in this compound is in the form $HRPFe^{IV} = O$, with an oxidation state above that of the native enzyme, which is a hemo-protein with a FeIII nucleus. When reacting with the phenolic substrate an intermediate complex of very low stability is obtained. From the complex, the native enzyme is again released, as well as the oxidation products, according to the following scheme:

$$HRPFe^{IV} = O + S \leftrightarrow [HRPFe^{IV} = O - S] \rightarrow HRPFe^{III} + P$$

where S is the phenolic substrate and P the reaction products. As regards to those reaction products, dimeric compounds can appear in the bulk reaction as a result of the binding of two of the radicals formed from the aromatic compound. Although the resulting dimer is less soluble, it remains in the aqueous solution where it can be oxidized again in successive steps, leading to the formation of a longer chain polymer whose precipitation is favoured. This requires additional consumption of hydrogen peroxide and influences the stoichiometry of the overall process, which moves towards gradually higher hydrogen peroxide/phenol molar ratios, depending on the degree of polymerization achieved. Although many works based on possible mechanisms have been found in the literature, only a few have proposed kinetic models and tried to solve the corresponding equations, fitting the model predictions with the experimental data. In this sense, it can be said that almost all the kinetic models developed and solved until now have been proposed by Nicell et al.

Models

Accepting the Chance-George mechanism and the deactivation phenomena previously described, Nicell proposed three kinetic models for the HRP/hydrogen peroxide/4-chlorophenol system, in the steady-state, fully transient and pseudo-steady-state, respectively [130]. As Nicell pointed out, the steady-state model led to an equation similar to the one obtained in a ping-pong bisubstrate enzymatic kinetic. The transient-state model was formulated as a series of nine simultaneous differential equations, containing ten kinetic parameters to be determined by fitting. Due to its complexity, the model must be solved by the Runge-Kutta method and required long computing times, although there may be problems of convergence if the step size is not correctly chosen when applying the mentioned numerical method. Despite these drawbacks, the model provided a good fit of the experimental data. The pseudo-steady-state used the steady-state approximation only with some reaction intermediates and, despite certain simplifications, it remains complicated and it is still necessary to use numerical methods and a long computation time. The fitting of the model to the experimental data is 2% less precise than with the fully transient model, but in the authors' opinion, this loss of precision can be justified in order to achieve simplicity and to reduce the computation time.

Later, Buchanan and Nicell [127] and Buchanan et al. [77] applied the above model equations to obtaining the design equations for plug flow and continuous tank reactors in steady-state. Good fitting was achieved between the experimental data and the equations obtained for the HRP/phenol system, the enzyme inactivation rate in the continuous tank being lower than the one calculated in the plug flow reactor, working with PEG excess and keeping the complexity grade of the models as explained previously.

Later, Buchanan and Nicell [131] produced a simplified model where the active species of the enzyme are gathered in the so-called "active enzyme", E_a, which reduces the number of constants and simplify the calculation. Choi et al. [132] proposed a modification of the Chance-George mechanism, for the HRP/pentachlorophenol (PCP) system. According to these authors, intermediate enzyme substrate complexes (EH_2O_2, E_1PCP and E_2PCP) are produced in the first three steps of the mechanism. These authors obtained the initial reaction rate by using the steady-state approximation, checking that oxidation of pentachlorophenol with HRP follows a bisusbstrate ping-pong mechanism. A phenol oxidation process by HRP involving intermediate complex formation has also been proposed by Gilabert et al. [133], according to the following scheme:

By using a quantitative initial rate analysis, these authors determined the values of the Michaelis-Menten constants for phenol and hydrogen peroxide, as well as the enzyme catalytic constant value.

Tong et al. [134] have also described the oxidation reaction of phenol, 4-chlorophenol and 3-chlorophenol, catalyzed by horseradish peroxidase, as a bisubstrate ping-pong mechanism, and calculated the kinetic constants for different hydrogen peroxide concentrations, demonstrating that an optimum peroxide concentration exists, above which peroxide becomes a reaction inhibitor. On the other hand, Wu et al. [78] developed a kinetic model for the phenol/horseradish peroxidase system in the presence of polyethylene glycol. The phenol oxidation was described by using a Michaelis-Menten bisubstrate equation.

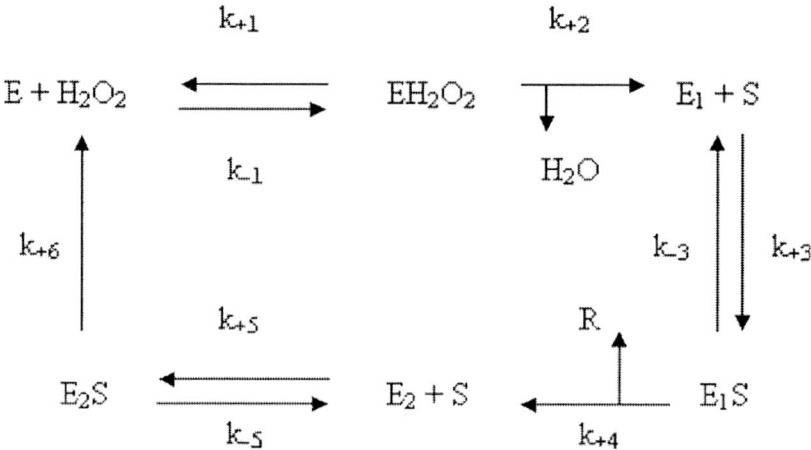

Although most kinetic studies have been carried out with horseradish peroxidase, some of them apply the Chance-George mechanism to the oxidation of phenols with soybean peroxidase [88, 90, 98, 135]. In this sense, Nicell and Wright [98] developed a model showing the dependence of the enzyme activity on the H_2O_2 concentration and they have also calculated the kinetic constants, which were lower than those obtained for HRP. They also noticed that soybean peroxidase, through the E_3 form, is more susceptible to permanent inactivation with hydrogen peroxide than HRP. Nissum et al. [135] found that, in the presence of hydrogen peroxide, the mechanisms of both HRP and SBP are very similar, although Compound I of SBP is less stable at neutral pH. According to the Chance-George mechanism, the stoichiometric peroxide/phenol ratio must be 1:2. However, some studies have been found in the literature that obtained different values for this relation. For example, Al-Kassim et al. [95] obtained values within the range 0.5-0.83. A result approaching unity was obtained in other works [78, 127]. Furthermore, Buchanan and Han [35], working with *Arthromyces ramosus* peroxidase obtained and average value of 1.22 at high enzyme and phenol concentrations and a value of 1.01 at low concentrations. These different stoichiometries observed might be attributed to differences in the average number of monomers in the precipitated polymer.

As regards the nature of polymers, five dimmers and a trimer were identified in the reaction solution by Yu et al. [136] working with the horseradish peroxidase/phenol system. They showed that three of these dimmers are also peroxidase substrates and can be oxidized like phenol, following first-order kinetics obtained by simplifying a bisubstrate model. The dimmers are *p*-phenoxyphenol, *p,p*-bisphenol and *o,o*-bisphenol, the oxidation rate of the first being higher than the phenol oxidation rate. According to these authors, the extra-consumption of hydrogen peroxide that takes place above the stoichiometric relationship can be explained by the oxidation of the mentioned compounds. Finally, a deactivation enzyme model, based on the attack of phenoxy radicals on the enzyme active centre following second order kinetics and depending on the PEG concentration, is also proposed in this work.

As it has been commented, almost all the authors have agreed that the removal of phenolic compounds in the presence of peroxidases follows a bisubstrate kinetic and a ping-pong mechanism. In this sense the research group leaded by Professor Bodalo developed several works with preliminary models.

Gomez et al. [137] developed a new method for the determination of the intrinsic parameters of reaction in processes involving a high initial reaction rate. The usefulness of this alternative, which consists of determining several sets of apparent parameters at different times and then extrapolating these to time zero, is demonstrated by the linear dependence obtained between the apparent parameters and the reaction time. The method calculated the values of the intrinsic parameters (enzyme specific activity and Michaelis-Menten constants of both substrates) for the system under study and was checked with experimental reaction rate data for the soybean peroxidase/phenol/ hydrogen peroxide system.

Using the above methodology the same authors [138] developed a kinetic model based on the Ping-Pong Bi-Bi mechanism and applied for the removal of 4-chlorophenol with soybean peroxidase and hydrogen peroxide. The developed model was firstly applied to the obtained results in a discontinuous tank reactor. To check the viability of such a mechanism they used their alternative method for determining the intrinsic parameters. The next step was to study the removal of the same compound, 4-chlorophenol, but in a continuous tank reactor.

In another work [115] a continuous tank reactor was used to study the 4-chlorophenol removal process using soybean peroxidase and hydrogen peroxide. A simplified steady-state design model for the continuous tank reactor was formulated and an expression for the steady-state conversion as a function of the operational conditions and of the intrinsic kinetic parameters was obtained. By using the values of the intrinsic kinetic parameters obtained for the discontinuous reactor the theoretical values of the steady-state conversion were calculated for all the experimental conditions assayed. The good approach between the experimental and calculated values of the steady-state conversion confirmed that the proposed design model and the kinetic law used, as well as the values of the intrinsic kinetic parameters, were useful for predicting the behavior of the continuous reactor.

After this first attempt to modelise the kinetics of the phenolic compounds with soybean peroxidase the same research group proposed an extended kinetic model applied to the removal of phenolic compounds in discontinuous and continuous reactors.

According to the original Chance-George mechanism and accepting the intermediate complex formation proposed by Choi et al. [132] and Gilabert et al. [133], Gomez et al. [118] proposed an extended version of the mentioned mechanism taking into account the following points:

- The overall process consists of an indetermined number of catalytic cycles that depend on the operational conditions.
- The initial cycle corresponds to the Dunford mechanism, in the enhanced version of Gilabert et al. [133], which takes into account the formation of enzyme-substrate intermediate complexes and produces monomer radicals that, by coupling, can form dimmers with several isomeric structures [136].
- The dimmers formed in the first step remain soluble and react in the same cycle as phenol, leading to the formation of higher oligomers with higher peroxide consumption. Although the different dimmers have different reactivities with the enzyme, only one average reaction rate for all of them is assumed.
- This cycle continues to run as long as there is hydrogen peroxide in the reaction medium, with the trimer, tetramer, etc, acting as the new substrate that takes part in the cycle, whose sequence remains the same as in the initial cycle.

The initial cycle for phenol and the one corresponding to the dimmers are described below:

Steps 1 & 2:

$$E + H_2O_2 \xleftrightarrow{K_1} EH_2O_2 \xrightarrow{k_2} EO + H_2O$$

Steps 3 & 4:

$$EO + \Phi H_2 \xleftrightarrow{K_3} EO\Phi H_2 \xrightarrow{k_4} {}^*EOH + {}^*\Phi H$$

Steps 5 & 6:

$${}^*EOH + \Phi H_2 \xleftrightarrow{K_5} {}^*EOH\Phi H_2 \xrightarrow{k_6} E + H_2O + {}^*\Phi H$$

Step 7:

$${}^*\Phi H + {}^*\Phi H \xrightarrow{k_7} H\Phi - \Phi H$$

Steps 8 & 9:

$$EO + H\Phi - \Phi H \xleftrightarrow{K_8} EOH\Phi - \Phi H \xrightarrow{k_9} {}^*EOH + {}^*\Phi - \Phi H$$

Steps 10 & 11:

$${}^*EOH + H\Phi - \Phi H \xleftrightarrow{K_{10}} {}^*EOHH\Phi - \Phi H \xrightarrow{k_{11}} E + H_2O + {}^*\Phi - \Phi H$$

Step 12:

$${}^*\Phi - \Phi H + {}^*\Phi - \Phi H \xrightarrow{k_{12}} H\Phi - \Phi - \Phi - \Phi H$$

From these equations it is possible to obtain the correspondent reaction rates for phenol, dimmer and hydrogen peroxide with the equation that calculates the concentration variation of active species of the enzyme in the reaction medium.

Following these premises they presented a model for the kinetic study of the immobilised SBP/phenol/hydrogen peroxide system that extends the peroxidase cycle to the reaction products, which permits to obtain a more general kinetic equation than the previously proposed. At time zero, the equation is reduced to a ping-pong bisubstrate kinetic, which agrees with previous studies. Using an initial rate procedure [137], the three parameters of the initial rate equation can be determined. In addition, the phenomenon of radical deactivation and/or enzyme-sequestration with the precipitated olygomers/polymers, widely described in the literature, was also modelled for the studied system, being the result of the gradual covering of the catalytic particles that contain the enzyme, which determines a progressive loss of activity. The deactivation model is included in the kinetic model.

Additionally, to check the validity of the generalized Ping Pong bisubstrate kinetics for the enzymatic reaction, obtained from an expanded version of the Dunford mechanism in the discontinuous tank reactor, a transient design model for removing phenol with immobilized soybean peroxidase and hydrogen peroxide in a continuous tank reactor assuming an ideal mixing flow was developed [118]. The experimental data of the phenol conversion at the reactor outlet were fitted to the model and the additional parameters were also calculated. The model was solved by numerical calculations and the experimental data were fitted to the model with a good degree of agreement between experimental and calculated values of the phenol conversion. This confirmed the validity of the design reactor model as well as the kinetic rate equations and intrinsic parameters used, despite the fact that they were obtained in assays carried out in a batch reactor.

The same authors [117] also studied the removal of phenol in fluidized bed reactor and developed a reactor model based on the experimental results that predicts the system's behavior both in steady and transient state. This model considers the fluidized bed reactor as a plug flow reactor in series with an ideal mixer and follows a kinetic law based on the observed external mass transfer resistances in order to work out the process rate. The good agreement obtained between the experimental and calculated values of phenol conversion demonstrates that the model is valid as a predictive model for using this reactor configuration.

Finally, this research group [119] also developed a model to the reactor configuration consisted in the combination of a continuous tank reactor followed by purification of the effluent in an ultrafiltration membrane module in series. This reactor configuration was applied to the removal of 4-chlorophenol by means of its oxidation with soybean peroxidase in the presence of hydrogen peroxide. The model assumes that the reaction between the phenolic compound and hydrogen peroxide follows the free radical mechanism proposed by Dunford, according to which phenol consumption rate law is a bisusbstrate Ping-Pong kinetics equation. The reaction products also follow the same reaction mechanism, which involves additional hydrogen peroxide consumption. The results obtained from comparing the model predictions and experimental results allowed the authors to affirm that both the kinetics law and the reactor design model were adequate for describing the behavior of the system.

Chapter 6

REMOVAL OF NON-PHENOLIC COMPOUNDS WITH SBP

REMOVAL OF PAHS

Kraus et al [139] tested the ability of soybean peroxidase to catalyze a variety of polycyclic aromatic hydrocarbons in the presence of water-miscible organic cosolvents. Optimal pH values were between 2.0 and 2.5, with SBP half-life of 120 h at pH 2.5. Anthracene and 9-methylanthracene attained more than 90% conversion in the presence of 50% (v/v) dimethylformamide. Anthraquinone was the main product of anthracene oxidation, while 9-methylanthracene produced anthraquinone and 9-methanol-9,10-dihydroanthracene. In both cases the resulting products are less toxic than the original compounds.

REMOVAL OF AROMATIC AMINES

Crude soybean peroxidase isolated from soybean seed hulls has been successfully applied to the removal of aromatic amines. Al-Ansari et al [140] worked with crude dry solid SBP to catalyze the oxidative polymerization of phenylenediamines in the presence of hydrogen peroxide. They worked out the optimum operation conditions: acidic pH, between 4.5 and 5.6; molar ratio H_2O_2: substrate from 1.5 to 2.0; Minimum SBP concentration required for 95% conversion of 1.0 mM substrate from 0.002 to 0.01 U/mL. Additionally, it was shown that the resulting polymers could be removed by using a surfactant, sodium dodecyl sulphate (SDS), with minimum SDS required for the removal of polymeric products determined to be between 0.2 and 0.25 mM. In this case, addition of polyethylene glycol (PEG) did not enhance the removal efficiency.

The same authors utilized crude SBP again to catalyze the removal of various aromatic compounds from synthetic and refinery wastewater, including aryldiamines [141] and obtaining conversions higher than 95%.

Some patents can be found in the literature about biocatalytic oxidation of different pollutants, including aromatic amines, using soybean peroxidase. U.S. Pat. No. 5,508.180 mentions both N-dealkylation and polymerization of aromatic amines, giving especific examples of N-dealkylation of N,N-dimethyl aromatic amines and polymerization of aniline [142]. Also, this work points out that in some cases there is no need of extraction of the peroxidase from the hull and instead it can be directly introduced to the reaction medium. U.S. Pat. No. 5,178.762 deals with the use of SBP to treat wastewaters containing aromatic amines, among other compounds, removing the contaminants or converting them to less hazardous or more easily removable forms. This patent also presents the alternative of directly using soybean peroxidase hulls packed in a column where the contaminated water and the hydrogen peroxide are passing over [143].

REMOVAL OF NITROAROMATICS

Nitroaromatics, even when they are phenolic or aniline compounds are not suitable for peroxidase catalysis. However, nitroaromatic compounds can be converted to their corresponding amines which can be oxidize with peroxidase enzymes. Mantha et al. [17] studied the oxidation of nitrobenzene and nitrotoluenes (o-, m- and p-) with a continuous two-stage treatment consisting in the zero-valent iron reduction of the nitroaromatic compounds to the corresponding aniline and after that soybean peroxidase treatment in a plug-flow reactor. They optimized different operational parameters such as pH at neutral value, peroxide/substrate molar ratio of 1.5, enzyme concentration (different values varying for the nitroaromatic compound) and alum concentration between 50 and 100 mg/L. The reaction time for the two steps was between 5 and 5.5 hours.

Following the same two steps treatment, Patapas et al. [18] published another paper related to the removal of dinitrotoluenes (2,4-dinitrotoluene and 2,6-dinitrotoluene) with ion and peroxidase-catalyzed oxidative polymerization. They used two peroxidases from different sources, *Arthoromyces ramosus* and soybean peroxidase. They concluded that soybean peroxidase was more resistant to mild acid conditions whereas *Arthoromyces ramosus* peroxidase was more stable in neutral solutions. Besides, soybean peroxidase was found to have a greater hydrogen peroxide demand, although less amount of enzyme was required for the experiments.

So the use of the soybean peroxidase enzyme looks like a suitable alternative biological method to remove these nitroaromatic compounds.

REMOVAL OF BENZOTHIAZOLES

An-Ansari [141] studied the removal of hazardous aqueous pollutants such as aryldiamines, aryldiols, 2-mercaptobenzothiazole and phenol from wastewaters with crude soybean peroxidase, isolated from soybean seed hulls, and in the presence of hydrogen peroxide.

Different experiments were carried out in order to optimize pH, hydrogen peroxide/substrate concentration ratio and soybean peroxidase concentration required to reach at least 95% of conversion of these pollutants in synthetic and refinery wastewaters. They concluded that the oxidation with soybean peroxidase was a suitable alternative for the removal of these compounds.

REMOVAL OF PAPER DYES

Other application of soybean peroxidase enzyme is dye decolorization. Knutson et al. [144] studied the removal of two of the most common paper dyes, the stilbene dye, Direct Yellow 11, and the methine dye, Basazol 46L, with two enzymes, soybean and horseradish peroxidases. Their results demonstrate that these two recalcitrant dyes can be effectively decolorized by enzymatic treatments by horseradish and soybean peroxidase. Soybean peroxidase obtained good removal efficiencies from pH 4.5 to 8.5. Although soybean peroxidase was effective in removing both dyes, SBP degrades quicker Basazol 46L than Direct Yellow 11.

Chapter 7

CONCLUSION

Soybean peroxidase (SBP) has proven its efficiency in the removal of many different pollutants present in industrial effluents: phenolic compounds, polycyclic aromatic hydrocarbons, aromatic amines, nitroaromatics, benzothiazoles and paper dyes. In addition to the advantages commonly attributed to enzymatic treatments, such as operation within a wide range of conditions, high specificity towards the targeted compounds and minimum environmental impact, peroxidases allow an easy separation of the reaction products resulting from the catalytic process. And, furthermore, SBP might be able to overcome some typical problems associated to enzymatic treatments: the high cost of the treatment and enzyme inactivation due to inappropriate reaction conditions. SBP can be easily extracted from soybean seed hulls, where it is present at high concentrations. And since the seed coat is a by-product of the soybean food industry, SBP has high potential of being a cost effective alternative compared to other peroxidase sources. Additionally, its enzymatic activity can be maintained over a wide range of temperature and pH, reducing the risk of enzyme deactivation. The direct use of soybean seed hulls without extraction has been already tested with promising results and further research in this area, especially regarding the use of real samples of contaminated effluents instead of synthetic ones, can be a first step for the development of cost effective and highly efficient processes for wastewater treatment at industrial scale.

REFERENCES

[1] Bryant, C. W., Avenell, J. J., Barkley, W. A. & Thut, R. N. (1992). The removal of chlorinated organic from conventional pulp and paper wastewater treatment systems. *Water Science and Technology.* 26, 417-425.

[2] Czaplicka, M. (2004). Sources and transformations of chlorophenols in the natural environment. *Science of the Total Environment.* 322, 21-39.

[3] Ahlborg, U. G. & Thunberg, T. M. (1980). Chlorinated phenols: occurrence, toxicity, metabolism, and environmental impact. *Critical Reviews in Toxicology.* 7, 1-35.

[4] Batoev, V. B., Nimatsyrenova, G. G., Dabalaeva, G. S. & Palitsyna, S. S. (2005). An assessment of contamination of the Selenga River Basin by chlorinated phenols. *Chemistry for Sustainable Development.* 13, 31-36.

[5] Gellman, I. (1988). Environmental effects of paper industry wastewater – an overview. *Water Science and Technology.* 20, 59-65.

[6] Garden, S. (1990). Baseline levels of absorbable organic halogen in treated wastewaters from bleached kraft pulp mill in Ontario. *Chemosphere.* 20, 1695-1700.

[7] Keith, L. H. & Telliard, W. A. (1979). Priority pollutants: a prospective view. *Environmental Science and Technology.* 13, 416-424.

[8] EC Decision 2455/2001/EC of the European Parliament and of the Council of November 20, 2001 establishing the list of priority substances in the field of water policy and amending Directive 2000/60/EC (L 331 of 15-12-2001).

[9] Agency for Toxic Substances and Disease Registry (ATSDR). (1995). Toxicological profile for Polycyclic Aromatic Hydrocarbons (PAHs). Atlanta, GA: U.S. Department of Health and Human Services, Public Health Service.

[10] Guerin, W. F. & Jones, G. E. (1988). Mineralization of phenanthrene by a Mycobacterium sp. *Applied and Environmental Microbiology.* 54, 937-944.

[11] Heitkamp, M. A., Freeman, J. P., Miller, D. W. & Cerniglia, C. E. (1988). Pyrene degradation by a Mycobacterium sp.: Identification of ring oxidation and ring fission products. *Applied and Environmental Microbiology.* 54, 2556-2565.

[12] Wammer, K. H. & Peters, C. A. (2005). Polycyclic aromatic hydrocarbon biodegradation rates: A structure-based study. *Environmental Science and Technology.* 39, 2571-2578.

[13] Shengquan, Y., Hui, W., Chaoua, Z. & Fu, H. (2008). Separation of carcinogenic aromatic amines in the food colourants plant wastewater treatment. *Desalination.* 222, 294-301.

[14] Benigni, R. & Passerini, L. (2002). Carcinogenicity of the aromatic amines: from structure–activity relationships to mechanisms of action and risk assessment. *Mutation Research.* 511, 191-206.

[15] Karim, Z. & Husain, Q. (2009). Redox-mediated oxidation and removal of aromatic amines from polluted water by partially purified bitter gourd (Momordica charantia) peroxidase. *International Biodeterioration & Biodegradation.* 63, 587-593.

[16] Agrawall, A. & Tratnyek, P. G. (1996). Reduction of Nitro Aromatic Compounds by Zero-Valent Iron Metal. *Environmental Science and Technology.* 30, 153-160.

[17] Mantha, R., Biswas, N., Taylor, K. E. & Bewtra, J. K. (2002). Removal of Nitroaromatics from Synthetic Wastewater Using Two-Step Zero-Valent Iron Reduction and Peroxidase-Catalyzed Oxidative Polymerization. *Water Environment Research.* 74, 280-287.

[18] Patapas, J., Mousa Al-Ansari, M., Taylor, K. E., Bewtra, J. K. & Biswas, N. (2007). Removal of dinitrotoluenes from water via reduction with iron and peroxidase-catalyzed oxidative polymerization: A comparison between Arthromyces ramosus peroxidase and soybean peroxidase. *Chemosphere.* 67, 1485-1491.

[19] Andreozzi, R., Caprio, V. & Marotta, R. (2001). Oxidation of benzothiazole, 2-mercaptobenzothiazole and 2-hydroxybenzothiazole in aqueous solution by means of H_2O_2/UV or photoassisted Fenton

systems. *Journal of Chemical Technology and Biotechnology.* 76, 196-202.
[20] Valdes, H., Murillo, F. A., Manoli, J. A. & Zaror, C. A. (2008). Heterogeneous catalytic ozonation of benzothiazole aqueous solution promoted by volcanic sand. *Journal of Hazardous Materials.* 153, 1036-1042.
[21] Gaja, M. A., Knapp, J. S. (1998). Removal of 2-mercaptobenzothiazole by activated sludge: a cautionary note. *Water Research.* 32, 3786-3789.
[22] Sarnalk, S. & Kanekar, P. (1995). Bioremediation of colour of methyl violet and phenol from a dye-industry waste effluent using *Pseudomonas* spp. Isolated from factory soil. *Journal of Applied Bacteriology.* 79, 459-469.
[23] Knutson, K. P. (2004). Enzymatic Biobleaching of Recalcitrant Paper Dyes. Institute of Paper Science and Technology at Georgia Institute of Technology.
[24] World Bank Group. (1995). Dye Manufacturing. Pollution Prevention and Abatement Handbook.
[25] Gomez, M., Matafonova, G., Gomez, J. L., Batoev, V. & Christofi, N. (2009). Comparison of alternative treatments for 4-chlorophenol removal from aqueous solutions: Use of free and immobilized soybean peroxidase and KrCl excilamp. *Journal of Hazardous Materials.* 169, 46-51.
[26] Chin, W. C., Mei, L. C., Jen, L.C. & Zong, H. C. (2000). Biodegradation of 2,4,6-trichlorophenol in the presence of primary substrate by immobilized pure culture bacteria. *Chemosphere.* 41, 1873-1879.
[27] Ghioureliotis, M. & Nicell, J. A. (1999). Assessment of soluble products of peroxidase-calalyzed polymerization of aqueous phenol. *Enzyme and Microbial Technology.* 25, 185-193.
[28] Gomez, J. L., Bodalo, A., Gomez, E., Bastida, J., Hidalgo, A. M., Gomez, M. (2008). A covered particle deactivation model and an expanded Dunford mechanism for the kinetic analysis of the immobilized SBP/phenol/hydrogen peroxide system. *Chemical Engineering Journal.* 138, 460-473.
[29] Chibata, I. (1978). Immobilized enzymes: Research and development. New York, USA: John Wiley & Sons Inc.
[30] Zaborsky, O. R. (1973). Immobilized Enzymes. Cleveland, Ohio, USA: CRC Press.

[31] Arroyo, M. (1998). Inmovilización de enzimas. Fundamentos, métodos y aplicaciones. *Ars Pharmaceutica.* 39, 23-39.
[32] Kennedy, J. F. & Cabral, J. M. S. (1983). Solid Phase Biochemistry. New York, USA: Scouten W.H., Ed., Wiley Pub.
[33] Hamid, M. & Rehman, K. (2009). Potential applications of peroxidases. *Food Chemistry.* 115, 1177-1186.
[34] Ryan, B. J., Carolan, N. & O'Fagain, C. (2006). Horseradish and soybean peroxidases: comparable tools for alternative niches? *Trends in Biotechnology.* 24, 355-363.
[35] Buchanan, I. D. & Han Y. S. (2000). Assessment of the potential of *Arthromyces Ramosus* peroxidase to remove phenol from industrial wastewaters. *Environmental Technology.* 21, 545-552.
[36] Villalobos, D. A. & Buchanan, I. D. (2002). Removal of aqueous phenol by *Arthromyces ramosus* peroxidase. *Journal of Environmental and Engineering Science.* 1, 65-73.
[37] Ikehata, K., Buchanan, I. D. & Smith, D. W. (2003). Treatment of oil refinery wastewater using crude *Coprinus cinereus* peroxidase and hydrogen peroxide. *Journal of Environmental Engineering and Science.* 2, 463-472.
[38] Masuda, M., Sakurai, A. & Sakakibara, M. (2001). Effect of reaction conditions on phenol removal by polymerization and precipitation using *Coprinus cinereus* peroxidase. *Enzyme and Microbial Technology.* 28, 295-300.
[39] Kauffmann, C., Petersen, B. R. & Bjerrum M. J. (1999). Enzymatic removal of phenols from aqueous solutions by *Coprinus cinereus* peroxidase and hydrogen peroxide. *Journal of Biotechnology.* 73, 71-74.
[40] Al-Kassim, L., Taylor, K. E., Nicell, J. A., Bewtra, J. K. & Biswas, N. (1994). Enzymatic removal of selected aromatic contaminants from wastewater by a fungal peroxidase from *Coprinus macrorhizus* in batch reactors. *Journal of Chemical Technology and Biotechnology.* 61, 179-182.
[41] Al-Kassim, L., Taylor, K. E., Bewtra, J. K. & Biswas, N. (1994). Optimization of phenol removal by a fungal peroxidase from *Coprinus macrorhizus* using batch, continuous, and discontinuous semibatch reactors. *Enzyme and Microbial Technology.* 16, 120-124.
[42] Ayala, M., Robledo, N. R., Lopez-Munguia, A. & Vazquez-Duhalt, R. (2000). Substrate Specificity and Ionization Potential in Chloroperoxidase-Catalyzed Oxidation of Diesel Fuel. *Environmental Science and Technology.* 34, 2804-2809.

[43] Pickard, M. A., Kadima, T. A. & Carmichael, R. D. (1991). Chloroperoxidase, a peroxidase with potential. *Journal of Industrial Microbiology and Biotechnology.* 7, 235-241.

[44] Casella, L., Poli, S., Gullotti, M., Selvaggini, C., Beringhelli, T. & Marchesini, A. (1994). The chloroperoxidase-catalyzed oxidation of phenols. Mechanisms, selectivity and characterization of enzyme-substrate complexes. *Biochemistry.* 33, 6377-6386.

[45] Aoun, S., Chebli, C. & Baloulene, M. (1998). Noncovalent immobilization of chloroperoxidase into talc: catalytic properties of a new biocatalyst. *Enzyme and Microbial Technology.* 23, 380-385.

[46] Goszczynski, S., Paszczynski, A., Pasti-Grigsby, M. B., Crawford, R. L. & Crawford, D. L. (1994). New Pathway for Degradation of Sulfonated Azo Dyes by Microbial Peroxidases of *Phanerochaete chrysosporium* and *Streptomyces chromofuscust. Journal of Bacteriology.* 176, 1339-1347.

[47] Michel JR, F. C., Balachandra, S., Grulke, E. A. & Adinarayana, C. (1991). Role of Manganese Peroxidases and Lignin Peroxidases of *Phanerochaete chrysosporium* in the Decolorization of Kraft Bleach Plant Effluent. *Applied and Environmental Microbiology.* 57, 2368-2375.

[48] Grabski, A. C., Rasmussen, J. K., Coleman, P. L. & Burgess, R. R. (1996). Immobilization of manganese peroxidase from *Lentinula edodes* on alkylaminated Emphaze™ AB 1 polymer for generation of Mn^{+3} as an oxidizing agent. *Applied Biochemistry and Biotechnology.* 60, 1-17.

[49] Grabski, A. C., Grimek, H. J. & Burgess, R. R. (2000). Immobilization of manganese peroxidase from *Lentinula edodes* and its biocatalytic generation of Mn-III-chelate as a chemical oxidant of chlorophenols. *Biotechnology and Bioengineering.* 60, 204-215.

[50] Aitken, M. D. & Irvine, R. L. (1989). Stability testing of ligninase and Mn-peroxidase from *Phanerochaete chrysosporium. Biotechnology and Bioengineering.* 34, 1251-1260.

[51] Bogan, B. W., Schoenike, B., Lamar, R. T. & Cullen, D. (1996). Manganese peroxidase mRNA and enzyme activity levels during bioremediation of polycyclic aromatic hydrocarbon-contaminated soil with *Phanerochaete chrysosporium. Applied and Environmental Microbiology.* 62, 2381-2386.

[52] Cameron, M. D., Timofeevski, S. & Aust, S. D. (2000). Enzymology of *Phanerochaete chrysosporium* with respect to the degradation of

recalcitrant compounds and xenobiotics. *Applied Microbiology and Biotechnology.* 54, 751-758.
[53] Scheibner, K. & Hofrichter, M. (1998). Conversion of aminonitrotoluenes by fungal manganese peroxidase. *Journal of Basic Microbiology.* 1, 51-59.
[54] Geng, Z., Rao, K. J., Bassi, A. S., Gijzen, M. & Krishnamoorthy, N. (2001). Investigation of biocatalytic properties of soybean seed hull peroxidase. *Catalysis Today.* 64, 233-238.
[55] Bakalovic, N., Passardi, F., Ioannidis, V., Cosio, C., Penel, C., Falquet, L. & Dunand, C. (2006). PeroxiBase: A class III plant peroxidase database. *Phytochemistry.* 67, 534-539.
[56] Hiraga, S., Sasaki, K., Ito, H., Ohashi, Y. & Matsui, H. (2001). A large family of Class III plant peroxidase. *Plant and Cell Physiology.* 42, 462-468.
[57] Klibanov, A. M., Alberty, B. N., Morris, E. D. & Felshin, L. M. (1980). Enzymatic removal of toxic phenols and anilines from waste waters. *Journal of Applied Biochemistry.* 2, 414-421.
[58] Klibanov, A. M. & Morris, E. D. (1981). Horseradish peroxidase for the removal of carcinogenic aromatic amines from water. *Enzyme and Microbial Technology.* 3, 119-122.
[59] Klibanov, A. M. (1982). Enzymatic removal of hazardous pollutants from industrial aqueous effluents. *Enzyme Engineering.* 6, 319-323.
[60] Klibanov, A. M., Tu T. M. & Scott K. P. (1983). Peroxidase-catalyzed removal of phenols from coal-conversion waste waters. *Science.* 221, 259-260.
[61] Cooper, V. A. & Nicell J. A. (1996). Removal of phenols from a foundry wastewater using horseradish peroxidase. *Water Research.* 30, 954-964.
[62] Tong, Z., Quingxiang, Z., Wilson, S. & Min Q. (1999). Study of the removal of 4-chlorophenol from wastewater using horseradish peroxidase. *Toxicological and Environmental Chemistry.* 71, 115-123.
[63] Wagner, M. & Nicell J. A. (2001). Treatment of a foul condensate from kraft pulping with horseradish peroxidase and hydrogen peroxide. *Water Research.* 35, 485-495.
[64] Wu, J., Taylor, K. E., Bewtra, J. K. & Biswas N. (1993). Optimization of the reaction conditions for enzymatic removal of phenol from wastewater in the presence of polyethylene glycol. *Water Research.* 12, 1701-1706.

[65] Davidenko, T. I., Oseychuk, O. V., Sevastyanov, O. V. & Romanouskaya, I. I. (2004). Peroxidase oxidation of phenols. *Applied biochemistry and Microbiology.* 40, 542-546.
[66] Tong, Z., Quingxiang, Z., Hui, H., Qin, L. & Yi, Z. (1997). Removal of toxic phenol and 4-chlorophenol from waste water by horseradish peroxidase. *Chemosphere.* 34, 893-903.
[67] Wagner, M. & Nicell J. A. (2003). Impact of the presence of solids on peroxidase catalyzed treatment of aqueous phenol. *Journal of Chemical Technology and Biotechnology.* 78, 694-702.
[68] Nakamoto, S. & Machida, N. (1992). Phenol removal from aqueous solutions by peroxidase-catalyzed reaction using additives. *Water Research.* 26, 49-54.
[69] Wagner, M. & Nicell, J. A. (2002). Impact of dissolved wastewater constituents on peroxidase-catalyzed treatment of phenol. *Journal of Chemical Technology and Biotechnology.* 77, 419-428.
[70] Nicell, J. A., Saadi, K. W. & Buchanan, I. D. (1995). Phenol polymerization and precipitation by horseradish peroxidase enzyme and an additive. *Bioresource and Technology.* 54, 5-16.
[71] Tatsumi, K., Ichikawa, H. & Wada, S. (1995). Removal of chlorophenols from wastewater by immobilized horseradish peroxidase. *Organohalogen Compounds.* 24, 37-42.
[72] Peralta-Zamora, P., Esposito, E., Pelegrini, R., Groto, R., Reyes, J. & Duran, N. (1998). Effluent treatment of pulp and paper, and textile industries using immobilised horseradish peroxidase. *Environmental Technology.* 19, 55-63.
[73] Hu, L. (1996). Study on the catalytic removal of phenols from aqueous solution by immobilized enzyme. *Huanjing Kexue.* 17, 57-60.
[74] Kim, G. & Moon S. (2005). Degradation of pentachlorophenol by an electroenzymatic method using immobilized peroxidase enzyme. *Korean Journal of Chemical Engineering.* 22, 52-60.
[75] Bodalo, A., Gomez, J. L., Gomez, E., Bastida, J. & Maximo, M. F. (2006). Comparison of commercial peroxidases for removing phenol from water solutions. *Chemosphere.* 63, 626-632.
[76] Nicell, J. A., Bewtra, J. K., Biswas, N., St. Pierre, C., & Taylor, K. E. (1993). Enzyme catalyzed polymerization and precipitation of aromatic compounds from aqueous solution. *Canadian Journal of Civil Engineering.* 20, 725-735.
[77] Buchanan, I. D., Nicell, J. A. & Wagner, M. (1998). Reactor models for horseradish peroxidase-catalyzed aromatic removal. *Journal of Environmental Engineering.* 124, 794-802.

[78] Wu, J., Taylor, K. E., Bewtra, J. K. & Biswas, N. (1999). Kinetic model-aided reactor design for peroxidase-catalyzed removal of phenol in the presence of polyethylene glycol. *Journal of Chemical Technology and Biotechnology.* 74, 519-526.

[79] Lopez-Molina, D., Hiner, A., Tudela, J., Garcia-Canovas, F. & Neptuno, J. (2003). Enzymatic removal of phenols from aqueous solution by artichoke (*Cynara acolymus* L.) extracts. *Enzyme and Microbial Technology.* 33, 738-742.

[80] Duarte-Vazquez, M. A., Garcia-Almendarez, B. E., Regalado, C., & Whitaker, J. R. (2001). Purification and properties of a neutral peroxidase from turnip (*Brassic napus* L. var. purple top white globe) roots. *Journal of Agriculture and Food Chemistry.* 49, 4450-4456.

[81] Duarte-Vazquez, M. A., Ortega-Tovar, M. A., Garcia-Almendarez, B. E., & Regalado, C. (2003). Removal of aqueous phenolic compounds from a model system by oxidative polymerization with turnip (*Brassica napus* L Var purple top white globe) peroxidase. *Journal of Chemical Technology and Biotechnology.* 78, 42-47.

[82] Cochrane, M. P. (1993). Observations of the germ aleurone of barley: Phenol oxidase and peroxidase activity. *Annals of Botany.* 73, 121-128.

[83] Adler, P. R., Arora, R., El Ghaouth, A., Glenn, D. M. & Solar, J. M. (1994). Bioremediation of phenolic compounds from water with plant root surface peroxidases. *Journal of Environmental Quality.* 23, 1113-1117.

[84] Gillikin, J. W. & Graham J. S. (1991). Purification and developmental analysis of the major anionic peroxidase from the seed coat of Glycine max. *Plant Physiology.* 96, 214-220.

[85] Nissum, M., Feis, A. & Smulevich, G. (1998). Characterization of soybean seed coat peroxidase: resonance Raman evidence for a structure-based classification of plant peroxidases. *Biospectroscopy.* 4, 355-364.

[86] Smulevich, G., Paoli, M., Burke, J. F., Sanders, S. A., Thorneley, R. N. F. & Smith, A. T. (1994). Characterization of recombinant horseradish peroxidase C and three site-directed mutants, F41V, F41W, and R38K, by resonance Raman spectroscopy. *Biochemistry.* 33, 7398-7407.

[87] Feis, A., Howes, B. D., Indiani, C. & Smulevich, G. (1998). Resonance Raman and electronic absorption spectra of horseradish peroxidase isozyme A2: evidence for a quantum-mixed spin species. *Journal of Raman Spectroscopy.* 29, 933-938.

[88] Flock, C., Bassi, A. & Gijzen M. (1999). Removal of aqueous phenol and 2-chlorophenol with purified soybean peroxidase and raw soybean hulls. *Journal of Chemical Technology and Biotechnology.* 74, 303-309.

[89] McEldoon, J. P. & Dordick, J. S. (1996). Unusual thermal stability of soybean peroxidase. *Biotechnology Progress.* 12, 555-558.

[90] Wright, H. & Nicell, J. A. (1999). Characterization of soybean peroxidase for wastewater treatment. *Bioresource Technology.* 70, 69-79.

[91] Amisha, J. K. & Behere, D. V. (2003). Activity, stability and conformational flexibility of seed coat soybean peroxidase. *Journal of Inorganic Biochemistry.* 94, 236-242.

[92] Caza, N., Bewtra, J. K., Biswas, N. & Taylor, K. E. (1999). Removal of phenolic compounds from synthetic wastewater using soybean peroxidase. *Water Research.* 13, 3012-3018.

[93] Kinsley, C. & Nicell, J. A. (2000). Treatment of aqueous phenol with soybean peroxidase in the presence of polyethylene glycol. *Bioresource Technology.* 73, 139-146.

[94] Al-Kassim, L., Taylor, K. E., Bewtra, J. K. & Biswas, N. (1993). Aromatics removal from water by Arthromyces ramosus peroxidase. In K. G. Welinder, S. K. Rasmussen, C. Penel, & H. Greppin (Eds.), Plant Peroxidases: *Biochemistry and Physiology.* (pp. 197-200). Universidad of Geneva.

[95] Al-Kassim, L., Taylor, K. E., Bewtra, J. K. & Biswas, N. (1993). Evaluation of the removal of aromatic and halogenated unsaturated hydrocarbons from synthetic wastewater by enzyme catalyzed polymerization. *Proc. 48th Purdue Industrial Waste Conference, Lewis Publishers, Chelsea, MI.* 413-420.

[96] Al-Kassim, L., Taylor, K. E., Bewtra, J.K. & Biswas, N. (1995). Optimization and preliminary cost analysis of soybean peroxidase-catalyzed removal of phenols from synthetic and industrial wastewater. *Proc. 45th Canadian Chemical Engineering Conference, Quebec.*

[97] McEldoon, J. P., Pokora, A. R. & Dordick, J. S. (1995). Lignin peroxidase-type activity of soybean peroxidase. *Enzyme and Microbial Technology.* 17, 359-365.

[98] Nicell, J. A. & Wright, H. (1997). A model of peroxidase activity with inhibition by hydrogen peroxide. *Enzyme and Microbial Technology.* 21, 302-310.

[99] Taylor, K. E., Bewtra, J. K. & Biswas, N. (1998). Enzymatic treatment of phenolic and other aromatic compounds in wastewaters. Proc. 71st Annu. Conf. & Exposition, *Water Environment Federation.* 3. 349-360.

[100] Wilberg, K., Assenhaimer, C. & Rubio, J. (2002). Removal of aqueous phenol catalyzed by a low purity soybean peroxidase. *Journal of Chemical Technology and Biotechnology.* 77, 851-857.

[101] Taylor, K. E., Al-Kassim, L., Bewtra, J. K., Biswas, N. & Taylor, J. (1996). Enzyme based wastewater treatment: removal of phenols by oxidative enzymes. In M. Moo-Young, W. A. Anderson, & A. M. Chakrabarty (Eds.), *Environmental Biotechnology: Principles and Applications* (pp. 524-532). Dordrecht, Kluwer Academic Publishers.

[102] Dec, J. & Bollag, J. M. (1994). Use of plant material for the decontamination of water polluted with phenols. *Biotechnology and Bioengineering.* 44, 1132-1139.

[103] Kennedy, K., Alemany, K. & Warith, M. (2002). Optimisation of soybean peroxidase treatment of 2,4-dichlorophenol. Water SA, 28, April 2002.

[104] Bodalo, A., Gomez, J. L., Gomez, E., Hidalgo, A. M., Gomez, M. & Yelo, A. M. (2006). Removal of 4-chlorophenol by soybean peroxidase and hydrogen peroxide in a discontinuous tank reactor. *Desalination.* 195, 51-59.

[105] Parra, M., Martinez-Ruiz, J., Tomas V., Martinez-Gutierrez, R., Garcia-Canovas, F. & Tudela, J. (2009). Optimization of the soybean peroxidase catalysed biodegradation of 2,4,6-trichlorophenol. *New biotechnology.* 255, September 2009. Doi:10.1016/j.nbt.2009. 06.514.

[106] Tatsumi, K., Wada, S. & Ichikawa, H. (1996). Removal of chlorophenols from wastewater by immobilized horseradish peroxidase. *Biotechnology and Bioengineering.* 51, 126-130.

[107] Wu, Y., Bewtra, J. K., Biswas, N. & Taylor, K. E. (1993). Removal of phenol derivatives from aqueous solution by enzymatic reaction with additives. *Proceedings of the 48th Purdue Industrial Waste Conference* 421-431.

[108] Wu, Y., Taylor, K. E., Biswas, N. & Bewtra, J.K. (1998). A model for the protective effect of additives on the activity of horseradish peroxidase in the removal of phenol. *Enzyme Microbiology and Technology.* 22, 15-322.

[109] Gomez, J. L., Bodalo, A., Gomez, E., Bastida, J., Hidalgo, A. M. & Gomez, M. (2006). Immobilization of peroxidases on glass beads: an

improved alternative for phenol removal. *Enzyme and Microbial Technology.* 39, 1016-1022.

[110] Bodalo, A., Bastida, J., Maximo, M. F., Montiel, M. C., Gomez, M. & Murcia, M. D. (2008). A comparative study of free and immobilized soybean and horseradish peroxidases for 4-chlorophenol removal: protective effects of immobilization. *Bioprocess and Biosystems Engineering.* 31, 587-593.

[111] Bassi, A., Geng, Z. & Gijzen, M. (2004). Enzymatic removal of phenol and chlorophenols using soybean seed hulls. *Engineering in Life Sciences.* 4, 125-130.

[112] Trivedi, U. J., Bassi, A. S. & Zhu J. X. (2006). Investigation of phenol removal using sol-gel/alginate immobilized soybean seed peroxidase. *The Canadian Journal of Chemical Engineering.* 84, 239-247.

[113] Gacche, R. N., Firdaus, Q. & Sagar A. D. (2003). Soybean (Glycine max L.) seed coat peroxidase immobilized on fibrous aromatic polyamide: A strategy for decreasing phenols from industrial wastewater. *Journal of Scientific and Industrial Research.* 62, 1090-1093.

[114] Magri, M., Miranda, M. & Cascone, O. (2005). Immobilized of soybean seed coat peroxidase on polyaniline: Synthetis optimization and catalytic properties. *Biocatalysis and Biotransformation.* 23, 339-346.

[115] Gómez, J. L., Gómez, E., Bastida, J., Hidalgo, A. M., Gómez, M. & Murcia, M. D. (2008). Experimental behaviour and design model of a continuous tank reactor for removing 4-chlorophenol with soybean peroxidase. *Chemical Engineering and Processing.* 47, 1786-1792.

[116] Trivedi, U. J., Bassi, A. S. & Zhu, J. X. (2006). Continuous enzymatic polymerization of phenol in a liquid-solid circulating fluidized bed. *Powder Technology.* 169, 61-70.

[117] Gómez, J. L., Bódalo, A., Gómez, E., Hidalgo, A. M., Gómez, M. & Murcia, M. D. (2007). Experimental behaviour and design model of a fluidized bed reactor with immobilized peroxidase for phenol removal. *Chemical Engineering Journal.* 127, 47-57.

[118] Gómez, J. L., Bódalo, A., Gómez, E., Hidalgo, A. M., Gómez, M. & Murcia, M. D. (2008). A transient design model of a continuous tank reactor for removing phenol with immobilized soybean peroxidase and hydrogen peroxide. *Chemical Engineering Journal.* 145, 142-148.

[119] Murcia, M. D., Gómez, G., Gómez, E., Bódalo, A., Gómez, J. L. & Hidalgo, A. M. (2009). Assessing combination treatment, enzymatic oxidation and ultrafiltration in a membrane bioreactor, for 4-

chlorophenol removal: Experimental and modelling. *Journal of Membrane Science.* 342, 198-207.
[120] Gijzen, M., Huystee, R. & Buzzell, R. I. (1993). Soybean seed coat peroxidase. A comparison of high-activity and low-activity and low-activity genotypes. *Plant Physiology.* 103, 1061-1066.
[121] Steevensz, A., Al-Ansari, M. M., Taylor, K. E., Bewtra, J. K. & Biswas, N. (2009). Comparison of soybean peroxidase with laccase in the removal of phenol from synthetic and refinery wastewater samples. *Journal of Chemical Technology and Biotechnology.* 84, 761-769.
[122] Dunford, H. B. (1976). On the function and mechanism of action of peroxidases. *Coordination Chemistry Reviews.* 9, 187-251.
[123] Job, B. & Dunford, H. B. (1976). Substituent effect on the oxidation of phenols and aromatic amines by horseradish peroxidase compound I. *European Journal of Biochemistry.* 66, 607-614.
[124] Adediran, A. & Lambeir, A. (1989). Kinetics of the reaction of compound II of horseradish peroxidase with hydrogen peroxide to form compound III. *Journal of Biochemistry.* 186, 571-576.
[125] Nakajima, R. & Yamazaki, I. (1987). The mechanism of oxyperoxidase formation from ferryl preoxidase and hydrogen peroxide. *Journal of Biological Chemistry.* 262, 2576-2581.
[126] Arnao, M. B., Acosta, M., del Rio, J. A. del, Varón, R. & García-Cánovas, F. (1990). A kinetic study on the suicide inactivation of peroxidase by hydrogen peroxide. *Biochimica et Biophysica Acta.* 1041, 43-47.
[127] Buchanan, I. D. & Nicell, J. A. (1997). Model development for horseradish peroxidase catalyzed removal of aqueous phenol. *Biotechnology and Bioengineering.* 54, 251-261.
[128] Baynton, K. J., Bewtra, J. K., Biswas, N. & Taylor, K. E. (1994). Inactivation of horseradish peroxidase by phenol and hydrogen peroxide: a kinetic investigation. *Biochimica et Biophysica Acta.* 1206, 272-278.
[129] Patel, P. K., Mondal, M. S., Modi, S. & Behere, D. V. (1997). Kinetic studies on the oxidation of phenols by the horseradish peroxidase compound II. *Biochimica et Biophysica Acta.* 1339, 79-87.
[130] Nicell, J. A. (1994). Kinetic of horseradish peroxidase-catalyzed polymerization and precipitation of aqueous 4-chlorophenol. *Journal of Chemical Technology and Biotechnology.* 60, 203-215.
[131] Buchanan, I. D. & Nicell, J. A. (1999). A simplified model of peroxidase-catalyzed phenol removal from aqueous solution. *Journal of Chemical Technology and Biotechnology.* 74, 669-674.

[132] Choi, Y. J., Chae, H. J. & Kim, E. Y. (1999). Steady-state oxidation model by horseradish peroxidase for the estimation of the non-inactivation zone in the enzymatic removal of pentachlorophenol. *Journal of Bioscience and Bioengineering.* 88, 368-373.
[133] Gilabert, M. A., Hiner, A. N. P., García-Ruiz, P. A., Tudela, J., García-Molina, F., Acosta, M., García-Cánovas, F. & Rodríguez-López, J. N. Differential substrate behaviour of phenol and aniline derivates during oxidation by horseradish peroxidase: kinetic evidence for two-step mechanism. *Biochimica et Biophysica Acta.* 1699, 235-243.
[134] Tong, Z., Qingxiang, Z., Hui, H., Quin, L. & Yin, Z. (1998). Kinetic study on the removal of toxic phenol and chlorophenol from wastewater by horseradish peroxidase. *Chemosphere.* 37, 1571-1577.
[135] Nissum, M., Schiodt, C. B. & Welinder, K. G. (2001). Reactions of soybean peroxidase and hydrogen peroxide pH 2.4-12.0, and veratryl alcohol at pH 2.4. *Biochimica et Biophysica Acta.* 1545, 339-348.
[136] Yu, J., Taylor, K. E., Zou, H., Biswas, N. & Bewtra, J. K. (1994). Phenol conversion and dimeric intermediates in horseradish peroxidase catalyzed phenol removal from water. *Environmental Science and Technology.* 28, 2154-2160.
[137] Gómez, J. L., Bódalo, A., Gómez, E., Hidalgo, A. M. & Gómez, M. (2005). A new method to estimate intrinsic parameters in the Ping-pong bisustrate kinetic: application to the oxipolymerization of phenol. *American Journal of Biochemistry and Biotechnology.* 1, 115-120.
[138] Bódalo, A., Gómez, J. L., Gómez, E., Hidalgo, A. M., Gómez, M. & Yelo, A. M. (2007). Elimination of 4-chlorophenol by soybean peroxidase and hydrogen peroxide: kinetic model and intrinsic parameters. *Biochemical Engineering Journal.* 34, 242-247.
[139] Kraus, J. J., Munir, I. Z., McEldoon, J. P., Clark, D. S. & Dordick, J. S. (1999). Oxidation of Polycyclic Aromatic Hydrocarbons Catalyzed by Soybean Peroxidase. *Applied Biochemistry and Biotechnology.* 80, 221-230.
[140] Al-Ansari, M. M., Steevensz, A., Al-Aasm, A., Taylor, K. E., Bewtra, J. K. & Biswas. N. (2009). Soybean peroxidase-catalyzed removal of phenylenediamines and benzenediols from water. *Enzyme and Microbial Technology.* 45, 253-260.
[141] Al-Ansari, M. M. (2008). Removal of various aromatic compounds from synthetic and refinery wastewater using soybean peroxidase. *Masters Abstracts International.* 47, 175.

[142] Johnson, M. A., Pokora, A. R. & Cyrus Jr, W. L. (1996). Biocatalytic oxidation using soybean peroxidase. *US Patent.* 5508180.
[143] Pokora, A. R. & Johnson, M. A. (1993). Soybean peroxidase treatment of contaminated substances. *US Patent.* 5178762.
[144] Knutson, K., Kirzan, S., Ragauskas, A. (2005). Enzymatic biobleaching of two recalcitrant paper dyes with horseradish and soybean peroxidase. *Biotechnology Letters.* 27, 753-758.

INDEX

A

absorption, 5, 50
absorption spectra, 50
acetone, 16
acetonitrile, 16
acid, 12, 16, 39
activity level, 47
additives, 6, 13, 16, 21, 49, 52
adhesives, 15
adsorption, vii, 9, 28
alternative treatments, 45
amines, vii, 1, 2, 4, 11, 12, 37, 38, 41, 44, 48, 54
aniline, 38, 55
aqueous solutions, 45, 46, 49
aromatic compounds, 3, 12, 15, 27, 38, 49, 52, 55
aromatic hydrocarbons, vii, 1, 2, 4, 37, 41
aromatic rings, 4
aromatics, vii, 1
assessment, 43, 44
atoms, 3
authors, 13, 16, 20, 23, 26, 28, 29, 30, 32, 35, 38

B

bacteria, 45
benzene, 3, 4, 55
bioaccumulation, 3, 5
biocatalysts, 24
biochemistry, 49
biodegradability, 3
biodegradation, 20, 44, 52
biological activity, 6
bioremediation, vii, 1, 2, 25, 47
biotechnology, 52
bisphenol, 19, 32
body fluid, 4
breathing, 4

C

carbon, vii
carcinogen, 5
catalysis, 38
catalyst, 2, 22, 24
catalytic activity, 16
catalytic properties, 16, 47, 53
cell membranes, 5
cellulose, 6
chlorine, 3
classification, 8, 50
coal, 3, 4, 48

coal tar, 4
coatings, vii, 15
color, iv
complexity, 30
compounds, vii, 1, 2, 3, 5, 7, 8, 11, 12, 13, 15, 19, 21, 22, 23, 25, 26, 27, 29, 32, 33, 37, 38, 39, 41, 48, 49, 50, 51, 52, 55
computation, 30
computing, 30
configuration, 13, 24, 26, 35
consumption, 29, 32, 33, 35
contamination, 3, 43
convergence, 30
copyright, iv
corrosion, 5
cost, 1, 15, 20, 21, 25, 41, 51
covalent bond, 9
covalent bonding, 9
covering, 34
crude oil, 4
culture, 45
cycles, 33

D

damages, iv
database, 48
decomposition, 3
defence, 12
degradation, 44, 47
Department of Health and Human Services, 4, 44
derivatives, 22, 23, 52
detergents, 25
developing countries, 1
differential equations, 29
dimethylformamide, 37
discharges, 1
dyes, vii, 1, 2, 4, 5, 6, 12, 39, 41, 56

E

effluent, 24, 35, 45
effluents, vii, viii, 1, 3, 15, 41, 48
electron, 11, 28
encapsulation, 9
entrapment, 16
environmental impact, vii, 1, 8, 41, 43
Environmental Protection Agency, 5
enzymatic activity, viii, 16, 41
enzyme immobilization, 13, 24
enzymes, vii, 1, 2, 7, 8, 9, 11, 21, 22, 23, 24, 25, 26, 38, 39, 45, 52
EPA, 3
ethanol, 16
European Parliament, 43
experimental condition, 23, 33
exposure, 4
extraction, viii, 15, 25, 38, 41

F

fission, 44
fixation, 6
flexibility, 51
fluidized bed, 22, 23, 26, 35, 53
food industry, viii, 1, 15, 41
formaldehyde, 15
free radicals, vii, 11
fungi, 12
fungus, 12

G

garbage, 4
gel, 22, 23, 26, 53
Georgia, 45
glycol, 16, 20, 21, 28, 31, 38, 48, 50, 51

H

half-life, 37

Index

halogen, 43
harmful effects, 4
heavy metals, 26
heme, 11
hybrid, 22, 23
hydrocarbons, 4, 51
hydrogen, vii, 4, 11, 16, 20, 22, 23, 24, 25, 27, 28, 29, 30, 31, 32, 33, 34, 35, 37, 38, 39, 45, 46, 48, 51, 52, 53, 54, 55
hydrogen peroxide, vii, 11, 16, 20, 22, 23, 24, 25, 27, 28, 29, 30, 31, 32, 33, 34, 35, 37, 38, 39, 45, 46, 48, 51, 52, 53, 54, 55
hydrolytic stability, 21

I

ideal, 35
immobilization, 8, 9, 13, 21, 22, 24, 47, 53
immobilized enzymes, 21, 22, 23
inclusion, 9
industrialization, vii
ingestion, 5
inhibition, 21, 51
inhibitor, 30
iron, 3, 38, 44
isozyme, 50

K

kinetic constants, 30, 31
kinetic model, 27, 29, 30, 32, 33, 34, 55
kinetic parameters, 30, 32
kinetic studies, 31
kinetics, 11, 32, 33, 34, 35

L

lakes, 4
lifetime, 21
lignin, 6, 12

linear dependence, 32
liver, 5
liver damage, 5

M

manganese, 12, 47, 48
manufacturing, vii, 4, 5, 6
matrix, 21
mechanical properties, 21
media, 23, 27
membranes, 5, 24
metabolism, 43
metals, 26
methanol, 16, 37
methodology, 32
microorganism, vii
mixing, 35
model system, 50
modelling, 54
molar ratios, 29
molecular weight, vii, 4, 11, 15, 17
molecules, 11
monolayer, 21
monomers, 31
Moon, 49
mRNA, 47
mutagen, 3

N

naphthalene, 4
natural resources, vii, 6
nerve, 5
nitrobenzene, 5, 38
nucleus, 29

O

oil, vii, 4, 46
oligomers, 16, 33
operating range, 20
optimization, 13, 53

organic chemicals, 4
organic compounds, 5, 26
organic solvents, 16
oxidation, vii, 1, 8, 11, 22, 24, 27, 28, 29, 30, 31, 32, 35, 37, 38, 39, 44, 47, 49, 53, 54, 55, 56
oxidation products, 29
oxidation rate, 32
ozonation, 45
ozone, 8

P

paints, vii
Parliament, 43
patents, 38
pathways, 28
PCP, 30
permission, iv
peroxide, vii, 8, 11, 16, 20, 22, 23, 24, 25, 27, 28, 29, 30, 31, 32, 33, 34, 35, 37, 38, 39, 45, 46, 48, 51, 52, 53, 54, 55
pharmaceuticals, vii, 5
phenol, 16, 19, 20, 21, 22, 23, 25, 26, 28, 29, 30, 31, 32, 33, 34, 35, 39, 45, 46, 48, 49, 50, 51, 52, 53, 54, 55
phenol oxidation, 30, 31, 32
phenolic resins, 15
photocatalysis, 8
photolysis, 8
pitch, 4
plants, 12
plastics, 3, 4, 5
polycyclic aromatic hydrocarbon, vii, 1, 2, 37, 41, 47
polymer, 25, 28, 29, 31, 47
polymeric products, 38
polymerization, vii, 29, 37, 38, 44, 45, 46, 49, 50, 51, 53, 54
polymers, vii, viii, 21, 22, 31, 34, 37
precipitation, 7, 11, 29, 46, 49, 54
properties, 2, 3, 16, 21, 47, 48, 50, 53
protective coating, 15

proteins, 25
pulp, 43, 49
purification, 15, 24, 25, 35
purity, 52

R

radiation, 6
radical mechanism, 35
radicals, vii, 11, 28, 29, 32, 33
Raman spectroscopy, 50
reaction mechanism, 35
reaction medium, 21, 33, 34, 38
reaction rate, 7, 28, 30, 32, 33, 34
reaction time, 32, 38
reactions, 1, 11, 21, 28
reagents, 24
recommendations, iv
recycling, vii, 6
regeneration, 26
Registry, 44
replacement, 15
requirements, 7
resins, vii, 3, 4, 15
resources, vii, 6
respect, 47
rights, iv
rings, 4
risk assessment, 44
room temperature, 16
rubber, 5

S

salts, 26
scaling, 23
seed, viii, 1, 15, 22, 25, 26, 37, 39, 41, 48, 50, 51, 53, 54
selectivity, 47
silane, 21
silica, 22, 23
skin, 4
sludge, 45

Index

sodium, 20, 37
sol-gel, 26, 53
solid waste, 20
solubility, 4, 11
solvents, 5, 16
soybeans, 20
species, 12, 27, 28, 30, 34, 50
spectroscopy, 50
spin, 50
steel, 3
stoichiometry, 29
strategy, 53
substrates, 5, 20, 22, 25, 28, 32
suicide, 54
surface area, 23
surfactant, 37
synthesis, 15

T

talc, 47
tar, 4
temperature, viii, 1, 3, 7, 13, 16, 23, 41
testing, 47
textiles, 5
thermal stability, 16, 51
thermostability, 1, 15
toluene, 4

toxic effect, 2
toxicity, vii, 3, 6, 16, 43
transformation, 27
transformations, 43
transport, 24
trends, 2

U

uniform, 22, 26

V

vacuum, 8
vulcanization, 5

W

waste, 19, 20, 45, 48, 49
waste water, 48, 49
wastewater, vii, viii, 1, 2, 3, 4, 5, 7, 9, 11, 12, 13, 15, 19, 21, 26, 38, 41, 43, 44, 46, 48, 49, 51, 52, 53, 54, 55
water policy, 43
World Bank, 45